Advancing with AutoCAD Release 14

for Windows NT and Windows 95

Other titles from Bob McFarlane

Beginning AutoCAD   ISBN 0 340 58571 4

Progressing with AutoCAD   ISBN 0 340 60173 6

Introducing 3D AutoCAD   ISBN 0 340 61456 0

Solid Modelling with AutoCAD   ISBN 0 340 63204 6

Starting with AutoCAD LT   ISBN 0 340 62543 0

Advancing with AutoCAD LT   ISBN 0 340 64579 2

3D Draughting using AutoCAD   ISBN 0 340 67782 1

Beginning AutoCAD R13 for Windows   ISBN 0 340 64572 5

Advancing with AutoCAD R13 for Windows   ISBN 0 340 69187 5

Modelling with AutoCAD R13 for Windows   ISBN 0 340 69251 0

Using AutoLISP with AutoCAD   ISBN 0 340 72016 6

Beginning AutoCAD R14 for Windows NT and Windows 95   ISBN 0 340 72017 4

Modelling with AutoCAD R14 for Windows NT and Windows 95   ISBN 0 340 73161 3

Advancing with AutoCAD Release 14

for Windows NT and Windows 95

Bob McFarlane
MSc, BSc, ARCST, MIMechE, MIEE, MIED, MILog

CAD Course Leader, Motherwell College,
AutoDESK Educational Developer

A member of the Hodder Headline Group
LONDON • SYDNEY • AUCKLAND
Copublished in North, Central and South America
by John Wiley & Sons Inc.
New York • Toronto

Dedication: to Liz with love

First published in Great Britain 1999 by
Arnold, a member of the Hodder Headline Group,
338 Euston Road, London NW1 3BH

http://www.arnoldpublishers.com

Copublished in North, Central and South America by
John Wiley & Sons Inc., 605 Third Avenue
New York, NY 10158-0012

British Library Cataloguing in Publication Data
A catalogue record for this book is available from the British Library

Library of Congress Cataloging-in-Publication Data
A catalog record for this book is available from the Library of Congress

ISBN 0 340 74053 1
ISBN 0 470 35238 8 (Wiley)

1 2 3 4 5 6 7 8 9 10

Commissioning Editor: Sian Jones
Project Manager: Robert Gray
Production Editor: Liz Gooster
Production Controller: Sarah Kett
Cover design: Stefan Brazzo

Produced by Gray Publishing, Tunbridge Wells, Kent
Printed and bound in Great Britain by The Bath Press, Bath
and The Edinburgh Press Ltd, Edinburgh

What do you think about this book? Or any other Arnold title?
Please send your comments to feedback.arnold@hodder.co.uk

Contents

Preface

This book is intended for the AutoCAD Release 14 user who wants to advance their CAD knowledge and skills. It is assumed that the reader is familiar with the basic R14 commands and can draw in two dimensions and add hatching, dimensions, text styles, etc. The ability to use icons and dialogue boxes is essential.

The aim of the book is to introduce Release 14 topics which are considered to be 'more advanced': attribute extraction, customization of linetypes, hatch patterns, shapes and menus, for instance. After working through the exercises and activities in the book, readers will be more proficient at using AutoCAD R14, have improved their draughting skills and increased their productivity rate.

This book will provide an ideal companion to my two other AutoCAD Release 14 books for Windows NT and Windows 95 and will provide excellent background material for any Release 14 user. The topics in the book will provide a sound grounding for several of the City & Guilds CAD schemes as well as for students studying the Higher National Certificate (HNC) and the Higher National Diploma (HND) courses in Computer Aided Draughting and Design (CADD). These two courses were pioneered and developed by the author at Motherwell College and are becoming increasing popular at colleges throughout Britain.

Using the book

The book is intended to be an interactive teaching aid – the user will learn by completing worked examples. Different methods of activating commands will be discussed as will the AutoCAD prompts. Dialogue boxes and icons will be displayed when considered relevant to the topic being discussed.

The following format has been adopted:

- Release 14 prompts will be in *italics*
- user responses will be in **bold** type
- the symbol <R> or <RETURN> will require the user to press the return/enter key
- a two-button mouse is assumed with the terms:
  a) pick: requires a left-click
  b) right-click: is obvious.

Saving completed exercises

Most CAD users save their work in a directory (which is now a folder in Release 14) and this practice should be continued. For convenience I will use a folder with the name **R14CUST** and this will be referred to continually throughout the book.

Chapter 1

The standard sheet

Most AutoCAD users will have their own standard sheet/prototype drawing with variables set to their own requirements. In this book all work will be completed on A3 size paper unless otherwise stated, and a new A3 template file will be created to cover all our draughting requirements.

To create this standard sheet:

1 Begin AutoCAD R14 and:
   prompt   Start Up dialogue box
   respond   1. pick **Use a Wizard**
              2. pick **Advanced Setup**
              3. pick **OK**
   prompt   Advanced Setup dialogue box
   respond   1. Units: Decimal with 0.00 Precision then Next
              2. Angle: Decimal with 0 Precision then Next
              3. Angle Measure: East 0 then Next
              4. Angle Direction: Counter-clockwise then Next
              5. Area: 420 width and 297 length then Next
              6. Title Block: No title block then Next
              7. Layout: No paper space capability then Done.

2 Set the grid to 10 and the snap to 5 and other drawing aids to your own requirements.

3 Create the following layers:

name	*colour*	*linetype*	*usage*
0	white	continuous	general (the default layer)
OUT	red	continuous	outlines
CL	green	center	centre lines
HID	colour 20	hidden	hidden detail
DIM	magenta	continuous	dimensions
TEXT	blue	continuous	text items
SECT	colour 74	continuous	hatching.

4 Other layers may be added as required.

5 With layer 0 current draw a rectangle from 0,0 to 420,297 to represent our drawing area.

6 Display the toolbars: Draw, Modify, Object Snap and any others you require. Specific toolbars will be displayed when required.

7 Set the GRIPS and PICKFIRST system variables to 0. This will disable both the grip effect and the automatic window effect.

8 Set a text style with:
   a) name: STDA3
   b) font name: romans.shx
   c) height: 0.

9 Set a dimension style with:
   a) name: STDA3 created from STANDARD
   b) Geometry: spacing: 10
       extension lines: 2.5
       origin offset: 2.5
       overall scale: 1
       arrowheads: both closed filled, size 3
       center: mark with size 3
   c) Format: user defined: OFF
       force lines inside: OFF
       fit: Best Fit
       horizontal justification: centered
       vertical justification: above
       text: inside horizontal OFF
                               outside horizontal ON
   d) Annotation: Units: Decimal with Precision 0.00
                               Trailing ON
                  Angles: Decimal with Precision 0.0
                               Trailing ON
                  Text: Style STDA3
                               Height 3
                               Gap 1
   e) Save to STDA3.

10 Menu bar with **File–Save As** and:
   prompt   *Save Drawing As dialogue box*
   respond   1. scroll and pick at Save as type: **Drawing Template File (*.dwt)**
               2. enter File name as: **STDA3**
               3. pick **Save**
   prompt   *Template Description dialogue box*
   respond   1. enter: My template file (or anything you want)
               2. pick OK.

11 Step 10 has saved the STDA3 template file in the AutoCAD R14 template folder. This is satisfactory but you can also save your template file in your R14CUST folder. This is achieved by repeating the File–Save As sequence then:
   a) scroll and pick the Drawing Template File (*.dwt)
   b) entering the file name as STDA3
   c) scroll and pick the R14CUST folder
   d) pick Save
   e) enter the save template description as before
   f) pick OK.

12 Save the standard sheet as a drawing with the menu bar sequence **File–Save As** and:
   prompt   *Save Drawing As dialogue box*
   respond   1. scroll and pick at Save as type: **AutoCAD R14 Drawing (*.dwg)**
               2. ensure **r14cust** folder is current
               3. enter File name as: **STDA3**
               4. pick **Save**.

13 The above steps have saved the standard sheet as:
    a) a template file in the AutoCAD R14 template folder
    b) a template file in the R14CUST folder
    c) a drawing file in the R14CUST folder.

14 The reason for saving the standard sheet as template files in two folders is really as a safeguard.

15 Saving the standard sheet as a drawing file will be useful, as certain topics require a drawing file rather than a template file.

Chapter 2

Attributes

Attributes are text items which are attached to blocks. This attribute information could include details about a component's cost, size, part number, material, etc. Attribute data can be extracted from a drawing and used as input to other applications, e.g. databases, spreadsheets, CNC systems, etc. A worked example will demonstrate how attributes are created in a drawing.

Basic information

A computer shop has in its window six different makes of computer each displaying a different software package. For easy reference, the shop owner has a drawing of the window layout and each computer is represented on the drawing by an icon containing information about the position in the window, the computer make, RAM size, cost and the software being displayed. This information is detailed in Fig. 2.1(a), and it is this data which is to be entered as attributes.

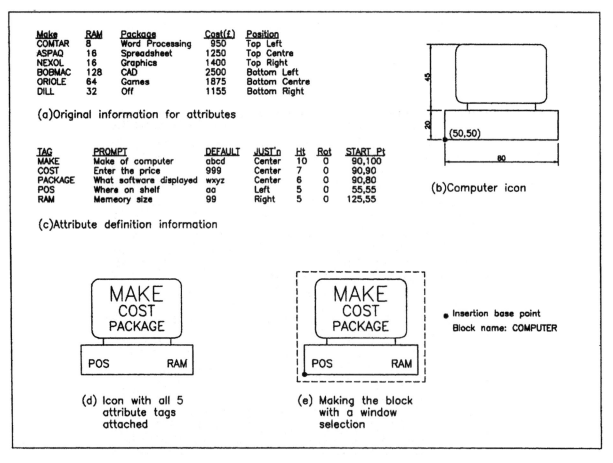

Make	RAM	Package	Cost(£)	Position
COMTAR	8	Word Processing	950	Top Left
ASPAQ	16	Spreadsheet	1250	Top Centre
NEXOL	16	Graphics	1400	Top Right
BOBMAC	128	CAD	2500	Bottom Left
ORIOLE	64	Games	1875	Bottom Centre
DILL	32	Off	1155	Bottom Right

(a) Original information for attributes

TAG	PROMPT	DEFAULT	JUST'n	Ht	Rot	START Pt
MAKE	Make of computer	abcd	Center	10	0	90,100
COST	Enter the price	999	Center	7	0	90,90
PACKAGE	What software displayed	wxyz	Center	6	0	90,80
POS	Where on shelf	aa	Left	5	0	55,55
RAM	Memeory size	99	Right	5	0	125,55

(c) Attribute definition information

(b) Computer icon

(d) Icon with all 5 attribute tags attached

(e) Making the block with a window selection

- Insertion base point
Block name: COMPUTER

Figure 2.1 Basic attribute information.

Making the icon

1 Open your STDA3 template file with layer OUT current.

2 Refer to Fig. 2.1(b) and draw the computer icon to the sizes given, using your discretion for sizes not specified. The only requirement is to position the lower left corner of the icon at the point 50,50. This is to assist with the position of the attributes.

Defining the attributes

1 Make layer TEXT current and refer to Fig. 2.1(c) which gives details about all the attributes which have to be added to the computer icon.

2 From the menu bar select **Draw–Block–Define Attributes** and:
    prompt   *Attribute Definition dialogue box*
    respond   *a*) enter **MAKE** in the Tag box
               b) pick the Prompt box and enter: **Make of computer**
               c) pick the Value box and enter: **abcd**
               d) alter Justification to **Center**
               e) ensure **STDA3** is the Text Style
               f) alter the height to **10**
               g) ensure the Rotation angle is **0**
               h) alter the Insertion Point to X:90, Y:100, Z:0
               i) dialogue box should resemble Fig. 2.2
               j) pick OK.

3 The attribute tag **MAKE** will be displayed in the computer icon, centred about the point 90,100.

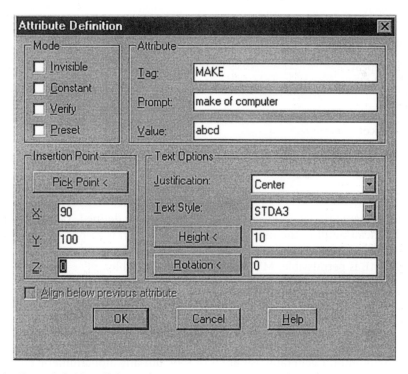

Figure 2.2 Attribute definition dialogue box.

4 At the command line enter **ATTDEF <R>** and:
    prompt   *Attribute modes-Invisible:N Constant:N.*
    enter   *(ICVP) to change, or press ENTER when done:*
    respond right-click
    prompt   *Attribute tag* and enter: **COST <R>**
    prompt   *Attribute prompt* and enter: **Enter the price <R>**
    prompt   *Default attribute value* and enter: **999 <R>**
    prompt   *Justification/Style/...*
    enter   **C <R>** – center text option
    prompt   *Center point* and enter: **90,90 <R>**
    prompt   *Height* and enter: **7 <R>**
    prompt   *Rotation angle* and enter: **0 <R>**.

5 The attribute tag **COST** will be displayed in the icon.

6 The details for the other three attributes (PACKAGE, POS, RAM) have still to be added to the icon. Using the Attribute Definition dialogue box or the ATTDEF command line entry, add these three attributes to the computer icon using the information given in Fig. 2.1(c).

7 When the five attributes have been defined, the computer icon will be displayed with the tags positioned as Fig. 2.1(d).

Making the block

When all the attributes have been defined, the computer symbol and tags have to be made into a block so:

1 At the command line enter **BLOCK <R>** and:
    prompt   *Block name* and enter: **COMPUTER <R>**
    prompt   *Insertion base point* and enter: **50,50 <R>**
    prompt   *Select objects*
    enter   **W <R>** – the window selection set option
    then   **window the icon and tags** as Fig. 2.1(e) then right-click.

2 The icon will disappear.

3 Remember OOPS?

4 *Note*: the block could have been created from a dialogue box with the menu bar sequence **Draw–Block–Make**.

Attribute terminology

When attributes are being defined, three words are used continually: *tag*, *prompt* and *value*:

Tag: this is the name (or label) given to the attribute being defined. Tag names are usually meaningful to the user and it is the tag which is displayed when the attribute definition is complete. Tag names should not have spaces, e.g. MY_NAME is permissible; MY NAME is not permissible.

Prompt: this should be a word or phrase which conveys a message to the user (who may not be you). When a block containing attributes is being inserted, the prompt will be displayed at the command line or in a dialogue box.

Value: this is a default for the attribute being defined and is displayed in **< >** brackets after the prompt. I generally use **abcd** or **999**, but any alphanumeric entry is permissible.

Inserting the block with attributes

There are two methods for inserting blocks containing attribute information, the method being controlled by the **ATTDIA** (attribute dialogue) system variable and:

ATTDIA: 0 – keyboard entry method

ATTDIA: 1 – dialogue box entry method

1   Still with the STDA3 template file on the screen which should be blank? If not, erase any objects from the screen.

2   Make layer OUT current and refer to Fig. 2.3.

3   At the command line enter **ATTDIA <R>** and:
    prompt   *New value for ATTDIA<?>*
    enter   **0 <R>**.

4   From the menu bar select **Insert–Block** and:
    prompt   *Insert dialogue box*
    respond   **pick Block**
    prompt   *Defined blocks dialogue box*, i.e. blocks in drawing
    respond   **pick COMPUTER then OK**
    prompt   *Insert dialogue box* with COMPUTER as block name
    respond   1. ensure Specify Parameters on Screen is on, i.e. tick
                 2. pick OK

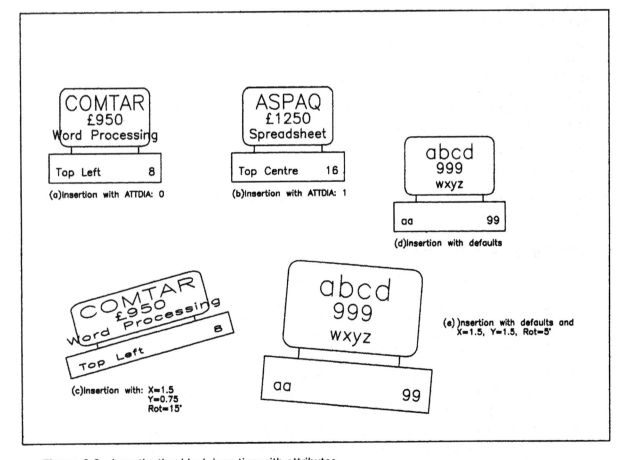

Figure 2.3 Investigating block insertion with attributes.

prompt	ghost image of computer icon
and	*Insertion point*
enter	**20,180 <R>**
prompt	*X scale factor* and enter: **1 <R>**
prompt	*Y scale factor* and enter: **1 <R>**
prompt	*Rotation angle* and enter: **0 <R>**
prompt	*Enter attribute values*
then	*Memory size<99>*, i.e. your prompt and default value
enter	**8 <R>**
prompt	*Where on shelf<aa>* and enter: **Top Left <R>**
prompt	*Make of computer<abcd>* and enter: **COMTAR <R>**
prompt	*Enter the price<999>* and enter: **£950 <R>**
prompt	*What software displayed<wxyz>*, enter: **SPREADSHEET <R>**.

5   The computer icon will be positioned as the entered insertion point and will display the attribute information added – Fig. (a).

6   *Note*:
   a) the order of your attribute prompts may differ from mine. This is normal when entering attributes from the keyboard. You should still be able to enter the correct information.
   b) The Specify Parameters on the Screen option in the Insert dialogue box is a toggle effect and when:
   ON (tick): insertion point is entered from the keyboard
   OFF (no tick): insertion point is entered from the dialogue box.
   c) The Insert dialogue box has two options:
   Block: to select blocks created in the current drawing
   File: to select other drawings (wblocks).

7   At the command line enter **ATTDIA <R>** and:

prompt	*New value for ATTDIA<0>*
enter	**1 <R>**.

8   Select the INSERT BLOCK icon from the Draw toolbar and:

prompt	*Insert dialogue box*
with	*COMPUTER as the block name*
respond	**pick OK**
prompt	*Insertion point* and enter: **150,180 <R>**
prompt	*X scale factor* and enter: **1 <R>**
prompt	*Y scale factor* and enter: **1 <R>**
prompt	*Rotation angle* and enter: **0 <R>**
prompt	*Enter Attributes dialogue box*
with	1. Block name: COMPUTER
	2. All prompts displayed
	3. All default values displayed for alteration
respond	alter the attribute default values pressing the <R> key after each entry

   a) Memory size   16
   b) Where on shelf   Top Centre
   c) Make of computer   ASPAQ
   d) Enter the price   £1250
   e) What software displayed   Spreadsheet
   f) dialogue box displayed as Fig. 2.4
   g) pick OK.

9   The computer block is inserted as Fig. (b).

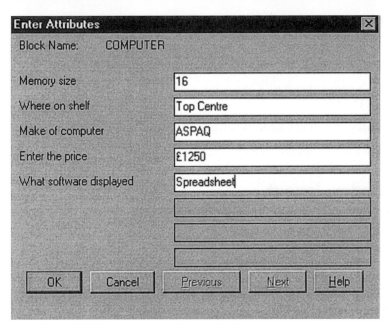

Figure 2.4 Enter Attributes dialogue box.

10 At the command line enter **INSERT <R>** and:
   prompt   Block name<COMPUTER> and right-click
   prompt   Insertion point and enter: **35,45 <R>**
   prompt   X scale factor and enter: **1.5 <R>**
   prompt   Y scale factor and enter: **0.75 <R>**
   prompt   Rotation angle and enter: **15 <R>**
   prompt   Enter Attributes dialogue box
   respond enter the same attribute values as step 4.

11 The block will be displayed as Fig. (c).

12 Repeat the INSERT command two more times with the following information:
Insertion point	265,145	170,30
X scale factor	1	1.5
Y scale factor	1	1.5
Rotation angle	0	−5
Attribute data	accept defaults	accept defaults
figure	(d)	(e).

Displaying attributes

Attribute data which has been added to inserted blocks can be 'turned off'. This is a useful option if the attribute information is not for general viewing.

1 Select the menu bar sequence **View–Display–Attribute Display–Off**.

2 The five icons will be displayed without the attribute data.

3 At the menu bar enter **ATTDISP <R>** and:
   prompt   Normal/ON/OFF/<Off>
   enter   **ON <R>**.

4 The attribute data is re-displayed.

Exploding blocks with attributes

1 Select the EXPLODE icon from the Modify toolbar and pick the inserted block at position (a). The block will be exploded and the original tag text will be displayed.

2 Repeat the explode command and pick the other inserted blocks. All the blocks will be exploded to display the tags.

3 Release 14 allows **all** inserted blocks to be exploded.

4 Now erase all the blocks but do not exit the drawing.

The shop window layout

1 Refer to Fig. 2.5 and create a window display using:
 a) the basic attribute data from Fig. 2.1(a)
 b) block insertion with $X=Y=1$
 c) the attribute entry method of your choice, i.e. command line or dialogue box.

2 Add any other refinements to the window layout.

3 When the six computer icons have been inserted with the attribute information added, save the layout as **R14CUST\WINDOW** as it will be used in other chapters.

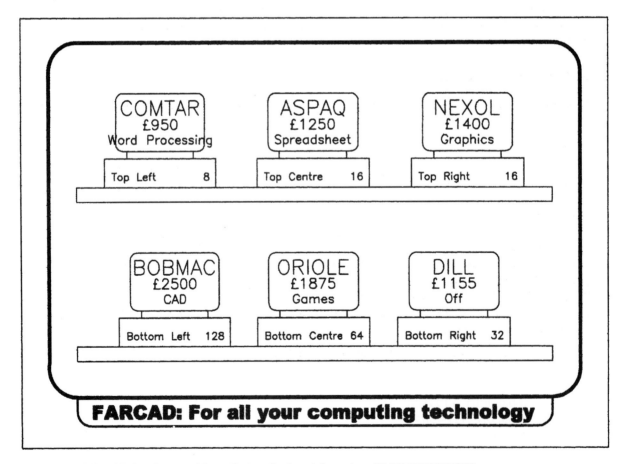

Figure 2.5   Window layout with attributes displayed (saved as R14CUST\WINDOW).

Summary

1 Attributes are text items attached to blocks and wblocks.

2 Blocks containing attributes are created and inserted in the usual manner.

3 Attributes are defined by a tag (label), a prompt and a default value.

4 The ATTDIA system variable allows attribute information to be entered:
   a) from the keyboard with ATTDIA 0
   b) from a dialogue box with ATTDIA 1.

5 Blocks with attributes can be inserted at varying X and Y scale factors and at varying rotation angles.

6 All blocks with attributes can be exploded to display the original tags.

7 Attributes attached to a block can be turned off with the ATTDISP command.

8 There are four attribute **modes**, these being:
   a) Invisible: attributes are not displayed (ATTDISP)
   b) Constant: attributes have a fixed value when inserted
   c) Verify: allows the user to verify the attribute value when inserted
   d) Preset: sets the attribute value to the default when inserted.

Assignment

Attribute examples usually involve a bit of keyboard entry, but this is normal. The example included for the attributes is the lorry park example which has been successful in my previous books.

Activity 1: Lorry park

Lorries being loaded at a warehouse have a different load and destination, and attributes have to be used to display this information.

1 Start with your STDA3 template file.

2 Refer to Activity 1(a) (all of the activity drawings are given at the end of the book) and:
   a) draw the lorry icon using sizes in Fig. (a)
   b) define the five attributes using the information in Fig. (b)
   c) when the attributes have been defined, the icon will be similar to Fig. (c).
   d) make a block of the icon and tags with the block name LORRY (the insertion point has been suggested for you and the actual layout is at your discretion)
   e) using the data from Fig. (d), insert the LORRY block and add the attribute data from Fig. (d). Use your imagination for the layout
   f) set the ATTDIA system variable to suit
   g) when all the blocks with attributes have been inserted the warehouse loading bay should resemble Activity 1(b).

3 When the layout is complete, save as **R14CUST\PARK**.

Chapter 3

Editing attributes

Attributes which are attached to blocks can be edited to correct errors or alter existing values. To demonstrate the editing of attributes, we will continue with our computer shop window layout.

The owner of the computer shop is concerned about his poor recent sales ever since a new superstore opened in a nearby business park, so he has decided to reduce the price of some of his computers in the window by having a sale. The computers reduced in price are:

COMTAR	£500
ASPAQ	£750
NEXOL	£850
DILL	£650 and displaying an Accounts software package.

Editing single attribute blocks

1 Open the window layout drawing R14CUST\WINDOW from the previous chapter.

2 Display the Modify II toolbar.

3 Select the EDIT ATTRIBUTE icon from the Modify II toolbar and:

 prompt   *Select block*
 respond **pick the COMTAR block**
 prompt   *Edit Attributes dialogue box* – looks familiar?
 with   entered attribute values displayed
 respond 1. alter price to **!! £500 !!**
            2. pick OK.

4 From the menu bar select **Modify–Object–Attribute–Single** and:
 a) select the ASPAQ block
 b) alter the price to !! £750 !!
 c) pick OK.

5 Using the icon or menu bar sequence, select the appropriate block and alter as follows:
 block   *price*   *software*
 NEXOL   !! £850 !!
 DILL   !! £650 !!   Accounts.

6 When the attributes have been edited, the window layout should resemble Fig. 3.1.

7 Do not exit the drawing.

Global attribute editing

Single attribute editing is very useful when only a few items require to be modified. If the drawing contains several attributes which require alteration, then global editing may be required. Attributes can be edited globally by specifying:
 a) the attribute block name
 b) the attribute tag
 c) a specific attribute value.

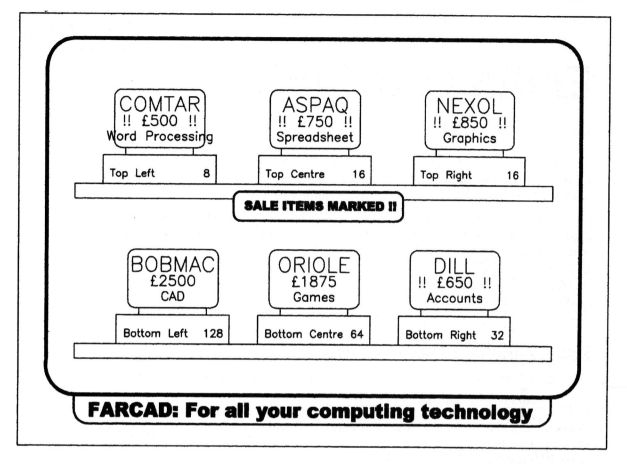

Figure 3.1 Window layout after single block attribute editing.

Editing by tag

The shop owner wants to the window layout to be more eye catching and has decided to change the text style of the computer name and price as well as the height and style of the RAM memory displayed on the icons. This means that several text styles have to be created.

1 Continue with the window layout example.

2 Menu bar with **Format–Text Style** and:

prompt	*Text Style dialogue box*
respond	**pick New**
prompt	*New Text Style dialogue box*
respond	**alter Style Name to ST1 then pick OK**
prompt	*Text Style dialogue box* with ST1 as Style Name
respond	1. alter Font Name to **Swis721BlkOulBT**
	2. alter height to 10
	3. pick Apply then Cancel.

3 Create another two text styles:

name	*font name*	*height*
ST2	Arial Black	8
ST3	Wide Latin	10.

4 Menu bar with **Modify–Object–Attribute–Global** and:
   prompt   *Edit attributes one at a time* and enter: **Y <R>**
   prompt   *Block name specification<*>* and right-click
   prompt   *Attribute tag specification<*>*
   enter   **MAKE <R>**
   prompt   *Attribute value specification<*>* and right-click
   prompt   *Select Attributes*
   enter   **W <R> and window the six blocks**
   prompt   *6 found*
   respond   right-click
   prompt   *Value/Position/Height/Angle/Style/Layer/Color/Next*
   and   an *X* appears at one of the MAKE values (mine was ASPAQ)
   enter   **S <R>** – the style option
   prompt   *Text style STDA3*
               New style or ENTER for no change
   enter   **ST1 <R>**
   prompt   *Value/Position/...*
   enter   **N <R>** – the next option and *X* moves to another MAKE value
   prompt   *Value/Position/...*
   enter   **S <R>** then **ST1 <R>**
   prompt   *Value/Position/...*
   respond   continue with the **N–S–ST1** entries until all the make value text styles have been changed to ST1.

5 At the command line enter **ATTEDIT <R>** and:
   prompt   *Edit attributes one at a time* and enter: **Y <R>**
   prompt   *Block name specification* and right-click
   prompt   *Attribute tag specification*
   enter   **RAM <R>**
   prompt   *Select Attributes*
   enter   **W <R> and window the six blocks then right-click**
   prompt   *Value/Position/...* and an *X* at a RAM value
   enter   **S <R>** then **ST3 <R>**
   prompt   *Value/Position/...*
   respond   continue with the **N–S–ST3** entries until all the RAM values have been altered to text style ST3.

6 Using the edit attribute command:
   a) enter the tag specification: COST
   b) enter W <R> and window the six blocks
   c) alter the text style to ST2.

Editing by value

A specific attribute value can be altered with the attribute edit command and we will demonstrate this by altering two of the window layout values:
a) the 16 RAM machines have been upgraded to 24 RAM
b) the Word Processing has been replaced with a Database package.

1 Altered window layout still displayed?

2 Activate the global edit attribute command (menu bar or command line) and:
   prompt   *Edit attributes one at a time* and enter: **Y <R>**
   prompt   *Block name specification* and right-click
   prompt   *Attribute tag specification* and right-click

prompt	*Attribute value specification*
enter	**16 <R>**
prompt	*Select attributes*
respond	enter **W <R> and window the six blocks right-click**
prompt	*2 found*
and	*and X at one of the 16 values*
prompt	*Value/Position/…* and enter: **V <R>**
prompt	*Change or replace* and enter: **R <R>**
prompt	*New attribute value*
enter	**24 <R>**
prompt	*Value/Position/…* and enter: **N <R>**
and	*X moves to the next 16 value*
enter	**V <R>** then **R <R>**
prompt	*New attribute value* and enter: **24 <R>**.

3 The command line will be returned, and the 16 RAM memory values will be altered to 24.

4 Using the edit attribute command:
   a) enter the attribute value: **Word Processing**
   b) window the six blocks
   c) activate the V(alue) and R(eplace) options
   d) enter **Database** as the new attribute value.

Editing by block

It is possible to edit attributes by specifying the block name at the block name specification prompt. As our example only uses the one block (COMPUTER), no editing has been undertaken with this option. You should be capable of block editing if required?

String attribute editing

It is possible to edit a specific 'string' within an attribute value by responding **N** to the first global edit prompt.

1 Menu bar with **Modify–Object–Attribute–Global** and:

prompt	*Edit attributes on at a time* and enter: **N <R>**
prompt	*Global edit of attribute values*
	Edit only attributes visible on the screen?<Y>
enter	**Y <R>**
prompt	*Block name specification* and right-click
prompt	*Attribute tag specification* and right-click
prompt	*Attribute value specification* and right-click
prompt	*Select attributes*
respond	1. enter W <R>
	2. window the six blocks
	3. right-click
prompt	*30 found*
then	*String to change*
enter	**Top <R>**
prompt	*New string*
enter	**T <R>**.

2 All blocks on the top shelf will be replaced with the attribute position as T Left, etc.

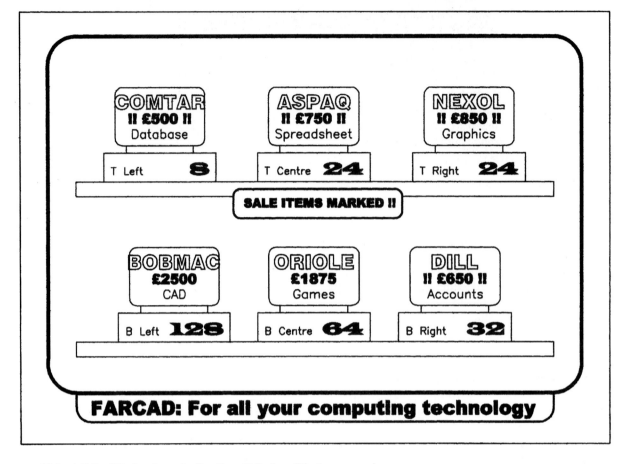

Figure 3.2 Window layout after the attribute editing commands.

3   Using the global attribute edit command:
  a) enter N to the first prompt
  b) right-click at the three specification prompts
  c) enter W and window the six blocks the right-click
  d) string to change: Bottom
  e) new string: B.

4   The editing exercise is now complete and the new window layout should resemble Fig. 3.2.

5   Save the layout at this stage as R14CUST\WINDMOD – **not window!**

DDedit

Most AutoCAD users should be familiar with the dynamic dialogue edit (DDEDIT) command as it is very useful for altering text displayed on the screen. The command also allows attribute tags to be altered.

1   Select the EDIT TEXT icon from the Modify II toolbar and:
  prompt   *<Select an annotation object>/Undo*
  respond **pick any attribute text item**
  and   nothing happens
  respond **right-click** to cancel the command.

2   With the EXPLODE icon from the Modify toolbar, select any block to display the attribute tags.

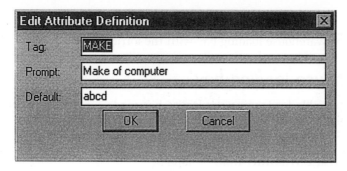

Figure 3.3 Edit Attributes dialogue box.

3 At the command line enter **DDEDIT <R>** and:
   prompt   *<Select an annotation object>*.
   respond **pick the MAKE tag from the exploded block**
   prompt   *Edit Attribute Definition dialogue box* as Fig. 3.3.

4 This dialogue box allows the attribute tag, prompt and default value to be edited if required.

5 Cancel the dialogue box then right-click.

6 The attribute editing exercise is now complete.

Summary

1 Single blocks with attributes can be edited using the EDIT ATTRIBUTE icon from the Modify II toolbar.

2 The ATTEDIT command and the Object–Attributes–Global selection allow attributes to be edited:
   a) by block
   b) by tag
   c) by value.

3 Global editing has several options and allows the value; position; height; angle; style; layer and colour of an attribute to be altered.

4 Global editing also allows individual strings in an attribute to be altered.

Assignment

This activity involves the warehouse loading bay with the lorries from Chapter 2.

Activity 2: Warehouse loading bay

The shipping clerk responsible for loading the lorries has made several mistakes, and these have to be corrected with attribute editing. This will involve both single and global editing.

1   Open the drawing file R14CUST\LORRY from Chapter 2.

2   Single editing:
   a) lorry A123 ROW has been re-routed to NEWCASTLE with COAL
   b) driver B. SLOW was held up in traffic and has been replaced with driver G.O. FAST
   c) no BOWLER HATS were available for the lorry to LONDON, and it was loaded with FERTILISER
   d) the lorries in bays 6 and 7 had their loads mixed up, and had to be emptied and re-loaded.

3   Global editing:
   a) all destinations should have the font name Dutch801XBdBT with a height of 8
   b) all loads have to be displayed with the font name City Blueprint at height of 6
   c) all drivers names have to be in red
   d) all registration numbers have to be in magenta
   e) the bay numbers have to be positioned at the front of the lorries with a height of 15.
   Note: two new text styles required!

4   String editing:
   a) all 2's have to be changed to TWO.

5   When all editing is complete, save the layout as R14CUST\PARKED.

Chapter 4

Extracting attribute data

Attributes which are attached to blocks in a drawing usually contain information which could be used with other software packages, e.g. databases and spreadsheets. AutoCAD allows attribute text information to be extracted from a drawing in three formats, these being:

1 Comma Delimited File (CDF): the extracted attribute data is separated with commas and quotes (") are placed around alphanumeric data but not around numeric data.

2 Space Delimited File (SDF): this extract file format has the data spaced out in field widths, alphanumeric data being positioned to the left of the field and numeric data positioned to the right.

3 Drawing Interchange File (DXF): this is the format most easily read by other CAD systems as well as CNC systems. The actual extract file is entirely different from the CDF and SDF files, and can be difficult to 'read' by the user.

Note

1 The file format to be used will depend on what software package the attribute data is to be used as input. Some packages prefer the CDF type format, while others require the data to be in fields – hence the SDF format.

2 The DXF format considered in this chapter should not be confused with the DXF export command which allows complete drawings to be exported. The DXF option with attributes is for block text data only.

Template file

Attribute data which is to be extracted from a drawing in CDF or SDF format requires a set of instructions on how and what data is to be extracted. These instructions are written by the user in a **template file** (an AutoCAD phrase) and the extracted attribute data has to be 'stored' in an extract file. When extracting attribute data, the user is therefore 'working' with three different files:

1 the drawing file containing the attribute information.

2 the template file to extract the attribute data.

3 the extract file to 'store' the extracted data.

The following points are considered important when dealing with template files:

1 a template file is a text file with the extension **.txt**.

2 a template file is written by the user with a text editor or any ASCII-compatible word processor.

3 In our exercises we will use the Notepad package.

4 An extract file is also a text file with the extension **.txt**. It is automatically created by AutoCAD but the user assigns the file name.

5 The general interaction between the drawing/template/extract files is shown in Fig. 4.1.

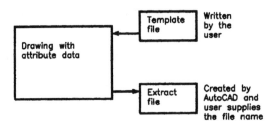

Figure 4.1 Interaction between the attribute drawing, the template file and the extract file.

Creating a template file

For our attribute extract exercise, we will return to the computer shop window layout. The shop owner wants a detailed list of the items in the window, and we will supply this as attribute data extract files. The template **file** has to be written, so:

1 Open the window layout drawing **R14CUST\WINDOW** from Chapter 2 to display the six computer icons with the original attribute data.

2 Left click on **Start** from the Windows taskbar then select **Programs–Accessories–Notepad** and:

    prompt   *Untitled Notepad screen*
    respond   enter the following lines as written
    ensure   1. the C and N entries **start at column 13**
              2. use the spacebar for spaces – **not TAB**
              3. press **<R>** at end of each line
              lines:

BL:NAME	**C008000**
BL:X	**N006001**
BL:Y	**N006001**
BL:LAYER	**C004000**
BL:HANDLE	**C004000**
MAKE	**C008000**
COST	**C008000**
PACKAGE	**C018000**
POS	**C018000**
RAM	**N004000.**

3 When all the lines have been written, select from the menu bar **File–Save As** and:

    prompt   *Save As dialogue box*
    respond   1. scroll and pick the **r14cust** folder
              2. enter File name as: **WINTEM.txt**
              3. pick **Save**.

4 Now minimize the Notepad screen by picking the leftmost icon from the title bar.

5 The window layout drawing screen will be returned and the Windows taskbar will display:

Start:AutoCAD–fig??:wintem–Notepad:any other current package.

Explanation of the template file

Before proceeding with the attribute extraction exercise, it is worthwhile explaining the layout of the template file as it will probably to new to most users.

1 A template file is a text file written by the user using any text editor and has the extension .txt.

2 It is recommended that the template and extract files are saved in the same folder as the drawing from which attributes have to be extracted – R14CUST in our example.

3 The template file should be meaningful to the user. Our template file had the name R14CUST\WINTEM.TXT and:
   a) R14CUST is the our folder name – a floppy could be used
   b) WINTEM.TXT is **MY** template file name created from:
      WIN – reference to the window layout drawing
      TEM – used to indicate a template file
      TXT – is the file extension.

4 The first five lines of the template file begin with **BL:**. This allows information about blocks to be extracted and:
   BL:NAME   the block name
   BL:*X*   the *X* coordinate of the block insertion point
   BL:*Y*   the *Y* coordinate of the block insertion point
   BL:LAYER   the layer of the inserted block
   BL:HANDLE   the inserted block handle number
   Other block information can be extracted, e.g. block rotation angle, *X* and *Y* scale factors, etc.

5 The rest of the lines in the template file contain references to the attributes which have to be extracted. The names used in these lines **must be identical to the attribute tags in the block definition**.

6 The C008000 and N006001 entries refer to the **type** of attribute data to be extracted and each entry is divided into three sections – C|008|000 and N|006|001 where:
   a) C: is for alphanumeric data, e.g. COMTAR, £950
   b) 008: is the 'length' of the data being extracted
   c) 000: all alphanumeric entries have 000 as the last three digits
   d) N: is for numeric data, e.g. 64, 165.0
   e) 006: is the numeric data 'length' and includes a decimal point
   f) 001: the numeric data will display one decimal place
   g) thus:
      C008000 is for character data with a field length of 8
      N006001 is for numeric data with a field length of 6 and with one decimal place.

7 To determine the field length, count the number of digits and/or letters of the largest **value** for a particular tag then add one or two extra digits.

8 When writing a template file, spaces should be entered with the spacebar – **never use the TAB key**.

9 The C and N entries **must begin at column 13**.

Extracting attribute data

Once the template file has been written and saved, it can be used to extract the attribute information from the blocks in the drawing.

1 Original window layout still displayed on screen?

2 At the command line enter **ATTEXT <R>** and:
   prompt   CDF, SDF or DXF Attribute extraction (or Objects)
   enter   **C <R>** – the CDF option
   prompt   Select Template File dialogue box
   respond 1. scroll and pick r14cust folder – probably is current
            2. pick **wintem**
            3. pick Open
   prompt   Create Extract File dialogue box
   respond 1. ensure r14cust name current
            2. enter File name as **wincdf.txt**
            3. pick Save.

3 If the template file has been written correctly then:
   prompt   6 records in extract file.

4 The extraction process has been successful, but nothing appears to have happened?

5 The SDF format can be extracted in a similar manner, but we will use the attribute extraction dialogue box, so at the command line enter **DDATTEXT <R>** and:
   prompt   Attribute Extraction dialogue box
   respond a) pick SDF format
            b) pick Template File
            c) scroll and pick r14cust folder
            d) pick wintem then Open
            e) pick Output File
            f) scroll and pick r14cust folder
            g) enter File name as **winsdf.txt** then Save
            h) pick Select Objects<
   prompt   Select objects
   enter   **W <R> and window the blocks then right-click**
   prompt   Attribute Extraction dialogue box similar to Fig. 4.2
   respond pick OK
   prompt   6 records in extract file.

6 The DXF format is extracted in a slightly different way to the CDF and SDF formats as the template file is not required. At the command line enter **ATTEXT <R>** and:
   prompt   CDF, SDF or DXF Attribute extraction
   enter   **D <R>** – the DXF option
   prompt   Create Extract File dialogue box
   respond 1. ensure r14cust folder is current
            2. enter File name as **windxf**
            3. pick save
   prompt   42(?) objects in extract file.
   and   drawing screen returned.

7 *Note*: a DXF attribute extraction file is not a **.txt** file. It is a text type file but has the extension **.dxx** which is a 'compiled' file.

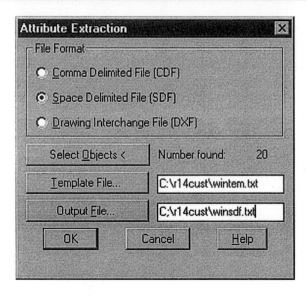

Figure 4.2 Attribute Extraction dialogue box.

Attribute extraction errors

While attribute extraction appears to be relatively simple, errors usually occur the first time that the ATTEXT command is used. These errors are usually in the template file and 'invalid field specification' and 'field overflow errors' are the usual messages displayed. The most common template file errors are:

a) not using the correct TAG names
b) using N for character fields instead of C
c) not specifying the correct field length
d) using the letter O instead of the number 0
e) not pressing <R> at the end of each line
f) having blank lines at the start or end of the file
g) using the TAB key instead of the spacebar.

Note: pre-R14 users note that the <R> key is pressed at the end of the last line. Have AutoDESK fixed this old problem?

Viewing extract files

Attribute extract files can be viewed and printed from Notepad and can also be imported into an AutoCAD drawing:

1   Window display still on the screen?

2   Left click on Notepad from the Windows taskbar to display the Notepad screen with the wintem template file.

    Note: if you did not minimize the Notepad screen, you will need to select Start–Programs–Accessories–Notepad.

3   From the Notepad menu bar select **File–Open** and:
    prompt   *Open dialogue box*
    respond   1. r14cust folder current?
               2. pick **wincdf**
               3. pick Open
    and   Notepad screen will display the wincdf.txt extract file.

4 Minimize the Notepad screen with the leftmost icon from the title bar. This will return the drawing screen and Notepad will be displayed in the Windows taskbar.

5 Make layer TEXT current and ensure that STDA3 is the current text style – this text style is important.

6 Select the MULTILINE TEXT icon from the Draw toolbar and:

prompt	*Specify first corner* and enter: **30,260 <R>**
prompt	*Specify opposite corner* and enter: **390,150 <R>**
prompt	*Multiline Text Editor dialogue box*
with	RomanS and height: 5 – the STDA3 text style
respond	**pick Import Text**
prompt	*Open dialogue box*
respond	1. scroll and pick r14cust folder
	2. pick **wincdf**
	3. pick Open
prompt	*Multiline Text Editor dialogue box*
with	the wincdf.txt file displayed
respond	pick OK.

7 The wincdf.txt extract file will be positioned on the drawing screen – Fig. 4.3(a). Note that I have ATTDISP set to OFF.

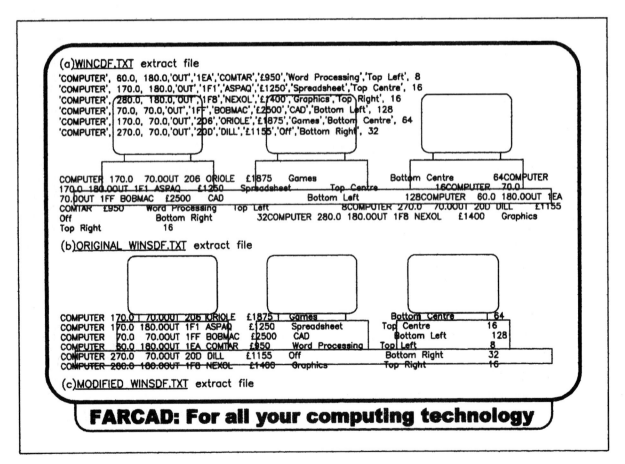

Figure 4.3 Window display with attribute text file imported (ATTDISP OFF).

8 At the command line enter **MTEXT <R>** and:
   a) specify first corner and enter: 30,190
   b) specify opposite corner and enter: 390,80
   c) pick Import Text
   d) pick file **r14cust\winsdf** then Open
   e) pick OK.

9 The winsdf.txt extract file will be displayed in the drawing as Fig. 4.3(b).

   Note: when I had completed this part of the exercise I was very surprised at the result of the winsdf.txt display. It was not as I had expected from pre-R14 SDF files. The nice structured column effect was missing. On investigating the original winsdf.txt file from Notepad, it would appear that all the attribute extract information has been grouped into a single string. I modified this original file and saved it as winsdfa.txt then imported it into the drawing as Fig. 4.3(c). The SDF files can be correctly opened in Word.

10 Activate Notepad from the Windows taskbar and:
   a) menu bar with File–Open
   b) alter Files of type: All Files
   c) pick windxf then Open
   d) scroll and study the screen – this is a dxf attribute extract file but is entirely different from the CDF and SDF formats
   e) menu bar with File–Exit.

11 The attribute extract exercise is now complete.

Summary

1 Attribute text data can be extracted from a drawing in three formats:
   a) Comma Delimited File – CDF
   b) Space Delimited File – SDF
   c) Drawing Interchange File – DXF.

2 The CDF and SDF extract formats are suitable for importing into other software packages such as databases and spreadsheets. The DXF format is suitable for other CAD systems.

3 Both the CDF and SDF formats require a template file to be used.

4 Template files must be written by the user using a text editor.

5 A template file is a text file.

6 The CDF and SDF extract files are text files with the .txt extension.

7 Extracting CDF and SDF attribute data involves three files:
   a) the drawing file (.dwg) with the attribute data
   b) the template file (.txt)
   c) an extract file (.txt).

8 The template and extract files should be stored in the same folder as the drawing file.

9 All text files can be imported into an AutoCAD drawing using the multiline (paragraph) text command.

Assignment

The attribute extraction activity will involve the original lorry warehouse loading bay from Activity 1.

Activity 3: Lorry loading bay

The original lorry attributes have to be extracted in CDF and SDF formats and then imported into the original drawing. As attribute extraction is a new concept, I have written the template file required for this activity.

The procedure is:

1   Open the R14CUST\PARK drawing from Activity 1.

2   Create the template file using Notepad.

3   Extract the attribute information in both CDF and SDF formats.

4   My suggested file names are:
    a) template file: r14cust\PARKTEM.txt
    b) CDF extract file: r14cust\PARKCDF.txt
    c) SDF extract file: r14cust\PARKSDF.txt.

5   When the CDF and SDF extract files have been created, import them into the original drawing using the MTEXT command.

6   When complete, save as R14CUST\PARKATT.

7   The template file is:

    BL:NAME   C006000
    BL:*X*   N006001
    BL:*Y*   N006001
    BL:ORIENT   N005001
    REG   C010000
    DEST   C012000
    DRIV   C010000
    LOAD   C014000
    BAY   N002000.

8   *Note*:
    a) BL:ORIENT is the block rotation angle
    b) I set ATTDISP to OFF when importing the extract files
    c) I also modified the PARKSDF extract file.

External references

Most users will be aware that a wblock can be inserted into any drawing and that drawings which contain wblocks are not automatically updated if one of the original wblocks is modified. External references (or **xrefs**) are similar to wblocks in that they can be inserted into a drawing, but they have one major advantage over wblocks. Drawings which contain external references are automatically updated if the original external reference 'drawing' is modified.

External references will be demonstrated by a worked example. The procedure will appear to be rather involved as it requires the user to open and save several times, but the final result is well worth the time and effort spent, so persevere with the exercise. For the demonstration we will:
a) create two simple external reference drawings
b) attach these two external references to three drawings
c) modify the two original external references
d) view the three original drawings
e) investigate the external references.

Creating the two external references.

1   Select from the menu bar **File–New** and:
    a) pick Use a Wizard with Quick Setup
    b) Decimal units with area 420 × 297
    c) set the grid to 10 and the snap to 5.

2   Make a new layer XREF1, colour blue and current.

3   Refer to Fig. 5.1 and draw a 20 unit square with both diagonals. Use a snap point as a corner – it will help later.

4   At the command line enter **BASE <R>** and:
    prompt   Base point<0,0,0>
    respond   **INTersection of the diagonals**.

5   Menu bar with **File–Save As** and save the square as R14CUST\XX1.

6   Menu bar with File–New and use the same entries as step 1, i.e. Use a Wizard with Quick Setup.

7   Make a new layer XREF2, colour green and current.

8   With the SNAP on, draw a polygon:
    a) with 6 sides
    b) centre at a snap point
    c) inscribed in a circle of radius 25.

9   Menu bar with **Draw–Block–Base** and:
    prompt   Base point<0,0,0>
    respond   **pick the polygon 'snap centre point'**.

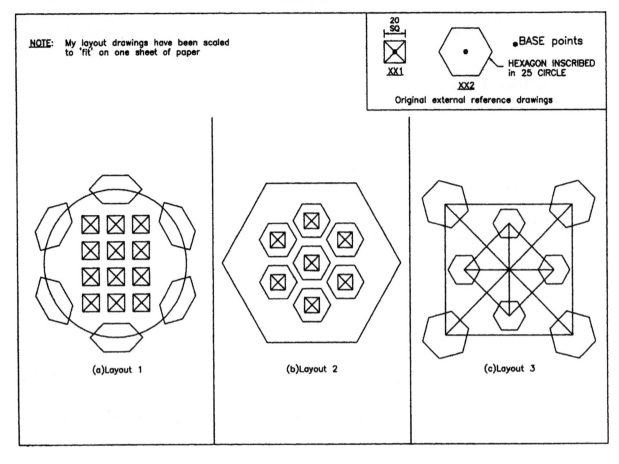

Figure 5.1 Layout drawings with original XREFS attached.

10   Menu bar with **File–Save As** and save the polygon as R14CUST\XX2.

11   *Note.* We have:
     a) created two drawings XX1 and XX2
     b) set a base point relative to the square and polygon to assist with the Xref insertion.

Drawing layout 1

1   Open your STDA3 template file and display the Draw, Modify, Object Snap and References toolbar.

2   With layer OUT current draw a circle, centre: 190,135, radius: 85.

3   Select the EXTERNAL REFERENCE ATTACH icon from the References toolbar and:

prompt	*Select file to attach dialogue box*
respond	1. scroll and pick r14cust folder
	2. pick xx1 drawing file
	3. pick Open
prompt	*Attach Xref dialogue box*
with	1. XX1 as the Xref Name
	2. Reference Type: Attachment
	3. *X,Y,Z* scale factors: 1
	4. Rotation angle: 0
	5. Specify On-screen, i.e. tick in box

respond   pick OK
prompt   *Attach Xref XX1: C:\r14cust\xx1.dwg*
            XX1 loaded
then   *Insertion point*
enter   **160,180 <R>**.

4   The blue square is inserted at the specified point.

5   Rectangular array the blue square:
    a) for 4 rows and 3 columns
    b) row distance: −30 and column distance: 30.

6   Select the External Reference Attach icon and:
    prompt   *Attach Xref dialogue box*
    respond   **pick Browse**
    prompt   *Select file to attach dialogue box*
    respond   1. pick r14cust\xx2 drawing file
                2. pick Open
    prompt   *Attach Xref dialogue box* with XX2 as Xref name
    respond   1. alter *X* scale factor to 1.25
                2. alter *Y* scale factor to 0.75
                3. pick OK
    prompt   *Attach Xref XX2: C:\r14cust\xx2.dwg XX2 loaded*
    then   *Insertion point*
    respond   **enter QUAD <R> and pick top of red circle**.

7   The green polygon will be inserted on the circle.
    Note: I found that the QUAD icon could not be activated at the Insertion point prompt!

8   Polar array the green circle with:
    a) red circle centre as the array centre point
    b) for 6 items, full circle with rotation.

9   Layout resembles Fig. 5.1(a).

10 Menu bar with File–Save As and enter R14CUST\LAYOUT1 as name.

Investigating the Xrefs

1   At the command line enter **BLOCK <R>** and:
    prompt   *Block name (or ?)* and enter: **? <R>**
    prompt   *Block(s) to list<*>* and right-click
    prompt   *AutoCAD Text Window*
    with   Defined blocks
                XX1   Xref: resolved
                XX2   Xref: resolved

User	External	Dependent	Unnamed
Blocks	References	Blocks	Blocks
0	2	0	0.

2   Cancel the text window.

3   Select the EXTERNAL REFERENCE icon from the References toolbar and:
    prompt   *External References dialogue box*
    with

Reference N	Status	Size	Type	Date saved	Path
XX1	Load	22	Attach	??	C:\r14cust\xx1.dwg

    respond   cancel the dialogue box.

4   Menu bar with Format–Layer and:
    prompt   Layer Properties dialogue box
    with   1. the STDA3 layers
               2. two new layers:
                    a) Xx1|xref1 blue
                    b) Xx2|xref2 green
    respond cancel the dialogue box.

5   *Note*: the new layers 'names' are different from normal. The (|) is a 'pipe symbol' and indicates that XX1 has been created on layer Xref1.

6   Now proceed to the second layout drawing – have you saved?

Drawing layout 2

1   Open the STDA3 template file with layer OUT current.

2   Draw a polygon with 6 sides and:
    a) centred on 190,135
    b) inscribed in a 120 radius circle.

3   Using the External Reference Attach icon, attach XX1 full size with 0 rotation at the following points:
    a) 145,165   *b*) 190,190   *c*) 235,165   *d*) 145,110   *e*) 190,135   *f*) 235,110   *g*) 190,80.

4   Attach external reference XX2 at the same seven points as step 3, full size with 0 rotation (snap on helps).

5   The result will resemble Fig. 5.1(b).

6   Menu bar with File–Save As and enter the name: R14CUST\LAYOUT2.

Drawing layout 3

1   Open the STDA3 template file, later OUT current.

2   Attach external references XX1 and XX2 in a layout of your own design. Fig. 5.1(c) is my layout.

3   Save this layout as R14CUSt\LAYOUT3.

Modifying the original Xrefs

1   Open drawing XX1 with layer XREF1 current.

2   Note the coordinates of the diagonal intersection then erase the blue lines – snap on helps.

3   Draw two concentric circles, centred on the noted coordinates with radii 10 and 20.

4   Menu bar with File–Save to automatically update XX1.

5   Open drawing XX2 with layer XREF2 current.

6   Note the coordinates on the 'polygon centre' (snap on) then erase the polygon.

7   Create a new text style with:
    a) name: XREF2
    b) font: Arial Rounded MT Bold with height: 15.

8   Draw the text phrase CAD, centred on the noted coordinates.

9   Menu bar with File–Save to automatically update XX2

Viewing the original layout drawings

1 Open drawing file LAYOUT1 and:
  a) note Preview displays original blue square and green polygon
  b) pick Open
  c) *prompt*:   *Resolve Xref XX1: c:\r14cust\xx1.dwg*
                     XX1 loaded
                     Resolve Xref XX2: c:\r14cust\xx2.dwg
                     XX2 loaded.

2 The layout 1 drawing will be displayed with the modified XX1 and XX2 external references, i.e. the blue circles and green text item as Fig. 5.2(a).

3 Check the Layer dialogue box – Xx1|xref1 and Xx2|xref2.

4 Select File–Save to update drawing Layout1.

5 Open drawing file Layout2 which will be displayed as Fig. 5.2(b) with the modified external references.

6 File–Save to update Layout2.

7 Finally open you Layout3 drawing which should be displayed with the modified external references – Fig. 5.2(c).

8 File–Save to update Layout3.

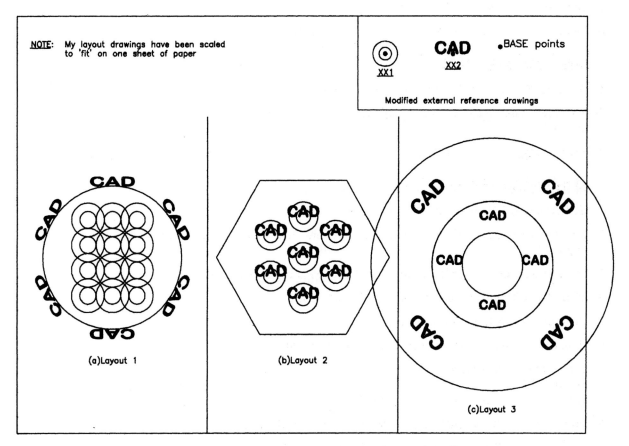

Figure 5.2 Layout drawings with modified XREFS.

Binding Xrefs

External references which are attached to a drawing are automatically updated when the original xref drawings are modified. This is due to the 'link' between the drawing and the attached xref. If this 'link is broken' then modifying the original xref will have no effect on the drawing layout.

1   Open the modified drawing Layout1.

2   Select the EXTERNAL REFERENCE icon and:
   prompt   *External References dialogue box*
   with   XX1 and XX2 displayed
   respond   1. **pick XX1** and other options available
               2. **pick Bind**
   prompt   *Bind dialogue box*
   respond   **Bind active the OK**
   prompt   *External References dialogue box*
   with   XX1 removed from list
   respond   1. pick XX2
               2. pick Bind
               3. pick OK from Bind dialogue box
               4. pick OK from External References dialogue box.

3   File–Save to update Layout1 drawing.

4   Open Layout3 drawing and repeat step 2 to bind the two external references XX1 and XX2.

5   File–Save to update layout2.

6   Before leaving this drawing, menu bar with Format–Layer and note the two layers:
   a) Xx1$0$xref1 and Xx2$0$xref2
   b) these indicate 'bound' layers, the (|) symbol being replaced by the (0) symbol.

Modifying the Xrefs again

1   Open drawing XX1 with layer XREF1 current and:
   a) note circle centre coordinates
   b) erase the circles
   c) draw a donut with ID: 15 and OD: 20, centred on the noted coordinates
   d) File–Save to update XX1.

2   Open drawing XX2 with layer XREF2 current and:
   a) note the coordinates of the text 'centre point'
   b) erase the text item
   c) draw an ellipse with centre at the noted coordinates and:
      i) axis endpoint: @20,0
      ii) other axis distance: @0,10
   d) File–Save to update XX2.

3   *a*) Open drawing Layout1 – circles and text displayed
   b) open drawing Layout3 – circles and text displayed
   c) open drawing layout2 – modified Xrefs (donuts and ellipses) should be displayed.

4   The **Bind** option 'breaks the link' with the external references and modifying the original Xrefs will not affect a 'bound' drawing.

5   This completes the external reference example.

Summary

1 External references are created in a manner similar to wblocks.

2 The BASE command is useful in defining an insertion base point.

3 External references are 'attached' to drawings and can be copied, arrayed, move, etc. as normal.

4 If an original external reference drawing is modified, **all drawings** containing the attached Xref are automatically updated to include this modification.

5 External references can be 'bound' to a drawing. When bound, they cease to be external references and are 'ordinary objects' and will not be updated if the original Xref is modified.

5 Layers created for external references:
a) display (|) if attached, e.g. Xx1|xref1
b) display (0) if bound, e.g. Xx1$0$xref1.

6 Realize that this has been a brief introduction to the concept of external references.

Assignment

An activity with external references is quite complex, but I have tried to include one which is relatively simple.

Activity 4: Flange arrangement

1 Begin a new drawing with Wizard–Quick Setup with decimal units and 420×297 size.

2 Create the original NUT using sizes given and set the BASE to the circle centre. Note the coordinates of circle centre for reference.

3 Save the drawing as NUT.

4 Open your STDA3 template file and draw three circles of radii 100, 30 and 65 (centre line).

5 Attach Xref NUT to top quadrant of 65 radius circle.

6 Polar array the attached Xref for 8 items.

7 Save the layout as XREFACT.

8 Open the original NUT drawing and modify as shown. Use your discretion for the slots but make the circle centre at the same coordinates as the previous circle.

9 File–Save to update NUT.

10 Open drawing XREFACT – modified NUT displayed?

Chapter 6

Customizing linetypes

AutoCAD Release 14 has several predefined linetypes which should be sufficient for most applications. There may, however, be the odd occasion when it is necessary to create new linetypes.

AutoCAD R14 allows four types of line to be created:
a) simple: consisting of dashes, dots and spaces, e.g. ____..____..____..
b) complex: with text items added, e.g. ____ME____ME____ME
c) complex: with shape items added, e.g. ____<->____<->____<->
d) multilines: consisting of several parallel elements, e.g. = = = = = = = = = = = =

In this chapter we will investigate how to create linetypes (*a*), (*b*) and (*d*) and leave the complex linetype (*c*) until we have investigated how to create shapes.

What are linetypes?

A linetype consists of dashes, spaces and dots, spaced out according to a predefined pattern. AutoCAD linetypes are contained in the **ACAD.LIN** or **ACADISO.LIN** files which are text files within the SUPPORT sub-folder in the AutoCAD R14 main folder. To view the existing linetype file:

1 Open your STDA3 template file.

2 Using the Windows taskbar, activate Notepad (i.e. Start–Programs–Accessories–Notepad) and:
   prompt   *Untitled Notepad screen*
   respond   menu bar with File–Open
   prompt   *Open dialogue box*
   respond   1. pick AutoCAD R14 – or your R14 folder name
               2. pick Support
               3. alter File name to: ***.LIN then <R>**
               4. pick acadiso
               5. pick Open.

3 The screen will display the AutoCAD ISO linetype file, two of these linetypes being:
   a) *BORDER, Border - - . - - . - - .
       A, 12.7, −6.35, 12.7, −6.35, 0, −6.35
   b) *CENTER, Center — – — – — -
       A, 31.75, −6.35, 6.35, −6.35.

4 Study the linetype file, then exit with File–Exit from the menu bar to return to the drawing screen.

Order when creating linetypes

When new linetypes are being created, the 'order' is:.

1 Create: by the user with a text editor in a .LIN file.

2 Load: from the .LIN file into AutoCAD.

3 Set: make new layers for the new linetypes.

4 Use: as appropriate.

Linetype descriptors

Every linetype in a .LIN file has a two line **DESCRIPTOR** consisting of three distinct parts:

1 The name of the linetype preceded by an asterisk (*), e.g. *CENTER. The name should be in CAPITALS.

2 A graphical description of the linetype which can be a text item or a pictorial representation of the linetype (or both). The pictorial representation need not be to scale and is made from dashes (_), dots (.) and spaces entered from the keyboard. It follows on from the linetype name, and is *separated from the name by a comma (,)*.

3 The actual linetype pattern line which is a coded definition of the linetype in a form which can be interpreted by a plotter, i.e. pen down and pen up movements. The actual values are **drawing units**. The coded definition for the CENTER linetype is A, 31.75, −6.35, 6.35, −6.35 which can be interpreted as:
   a) the **A**, is an 'alignment field' and is **ESSENTIAL**, i.e. every second line must begin with A
   b) 31.75 means a pen down movement of 31.75 drawing units
   c) −6.35 means a pen up movement of 6.35 drawing units
   d) 6.35 is a pen down movement of 6.35 drawing units
   e) −6.35 is a pen up movement of 6.35 drawing units.

Notes on linetype descriptors

1 A positive pattern line number signifies pen down, i.e. a line.

2 A negative pattern line number means pen up, i.e. no line (a space).

3 A zero pattern line number gives a dot.

4 A drawing unit is 1 mm.

5 The linetype 'appearance' is controlled by the LTSCALE variable and the ltScale option of the CHANGE command.

6 The linetype descriptor has a strict format and the syntax is important. The format **cannot be altered** and is:
   *NAME, dashes–spaces–dots
   A,pen down and pen up movements.

7 The actual number of pen movements in the second line need only be specified until the **REPEAT PATTERN** is completely defined, i.e. 31.75, −6.35, 6.35, −6.35 is the complete repeat pattern for the CENTER linetype.

8   The dash–space–dot description in the first line can be replaced with a written text item, e.g. Use for centre lines. The text **should not exceed 47 characters**. The comma is essential between the linetype name and the written description. If no written description is entered, then the comma is omitted.

9   The pen down and pen up movements **cannot exceed 12 in total**.

10 Figure 6.1(A) details the two linetype (BORDER and CENTER) from the ACADISO.LIN file and displays:
    a) the linetype repeat pattern with sizes as drawing units
    b) the linetype descriptors.

Creating two new simple linetypes

Linetypes are stored in text files with the extension **.LIN** and AutoCAD allows new linetypes to be created:
a) within the existing ACAD.LIN or ACADISO.LIN files
b) in user-defined .LIN files.

In our investigation we will create all our linetypes in a new file which we will call MYLINE. This will ensure that the two original .LIN files remain untouched – a wise precaution?

There are two methods for creating new linetypes:
a) from within AutoCAD using the LINETYPE command
b) by using a text editor (Notepad) either externally of from within AutoCAD.

We will create a linetype by each method. Fig. 6.1(B) displays these linetypes, detailing:
a) the linetype repeat pattern with sizes as drawing units
b) the complete linetype descriptors.

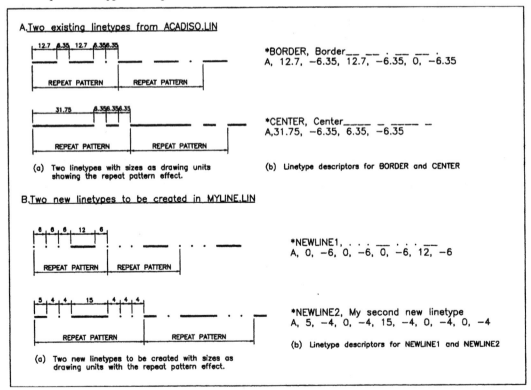

Figure 6.1   Linetype definitions and descriptors.

Using the LINETYPE command

1 At the command line enter **–LINETYPE <R>** and:
  prompt   ?/Create/Load/Set
  enter   **C <R>** – the create option
  prompt   Name of linetype to create
  enter   **NEWLINE1 <R>**
  prompt   Create or Append Linetype File dialogue box
  respond   1. scroll and pick the r14cust folder
          2. enter File name as **MYLINE**
          3. pick Save
  prompt   Creating new file
  then   Descriptive text
  enter   . . . _____ . . . _____ . . . _____ **<R>**
  prompt   A,
  enter   **0,–6,0,–6,0,–6,12,–6 <R>**
  prompt   New definition written to file
  then   ?/Create/Load/Set
  respond   **right-click** to end the command.

2 *Note*: the command line entry **-LINETYPE** will allow entries to be made from the command line. I believe this is easier for the user to understand at present.

Using a text editor (Notepad)

1 Activate Notepad from the Windows taskbar and:
  prompt   Untitled Notepad screen
  respond   1. menu bar with File–Open
          2. scroll and pick r14cust folder
          3. alter File name to ***.lin then <R>**
          4. pick MYLINE then Open
  prompt   :: [CODE_PAGE]—
  and   NEWLINE1 definition
  respond   1. move cursor below last line of text
          2. enter the following lines:
          *NEWLINE2, My second new linetype <R>
          A,5,–4,0,–4,15,–4,0,–4,0,–4 <R>

2 From the menu bar select File–Save to automatically update the r14cust\MYLINE.lin file.

3 Return to the drawing screen with File–Exit.

Loading new linetypes

Created linetypes have to be loaded before they can be used, either from the Layer Properties dialogue box, or using the command line entry.

1   At the command line enter **–LINETYPE <R>** and:

prompt	*?/Create/Load/Set*
enter	**L <R>** – the load option
prompt	*Linetype(s) to load*
enter	*** <R>** – the * wildcard for all linetypes
prompt	*Select Linetype File dialogue box*
respond	1. scroll and pick r14cust folder
	2. pick MYLINE then Open
prompt	*Linetype NEWLINE1 loaded*
	Linetype NEWLINE2 loaded
then	*?/Create/Load/Set*
and	right-click to end the command.

2   Enter -LINETYPE <R> at the command line and:

prompt	*?/Create/Load/Set*
enter	**? <R>** – the 'query' option
prompt	*Select Linetype File dialogue box*
with	r14cust and MYLINE displayed?
respond	pick MYLINE then Open
prompt	*AutoCAD Text Window*
with	*Linetypes defined in file c:\r14cust\MYLINE.lin*
	Name Description
	NEWLINE1 . . . ___ . . . ___ . . . ___
	NEWLINE2 My second new linetype
respond	1. cancel the text window
	2. right-click to end the -LINETYPE command.

3   *Note.*

If there are any errors in the created linetypes a message will be displayed. The most common error messages are:

a) there must be between 2 and 12 dash/dot spaces
b) commas must separate dash/dot specs
c) bad linetype pattern
d) dots (zeros) cannot be adjacent.

Errors are corrected by opening the .lin file in Notepad, or by re-creating the linetype with the -LINETYPE command.

Setting and using new linetypes

When a new linetype has been created and loaded it should be 'set' to a named layer (as all good CAD users know?) and then used to draw objects.

The computer shop owner was last encountered having a sale. This sale resulted in all stock being sold, and the shop owner moved his premises to a large retail park in a new redeveloped site (also where the lorry warehouse is situated – coincidence?). It is this retail park which will be drawn using the new created linetypes and new layers will be created for these linetypes, so:

1 Still with your 'blank' STDA3 template file on the screen?

2 With Format–Layer make two new layers using:

name	linetype	colour	usage
L1	NEWLINE1	number 20	shop outlines
L2	NEWLINE2	number 200	car park lines.

3 Note that when the two new linetypes are being 'set' to the new layers, they are displayed in the Select Linetype dialogue box.

4 Refer to Fig. 6.2 and design a layout of your choice with:
  a) layer L1 (NEWLINE1) for the shop outlines
  b) layer L2 (NEWLINE2) for the car park lines
  c) add suitable text
  d) note that my Fig. 6.2 displays other linetypes. Do not try to create or draw these yet
    – they are the complex linetypes from the next section.

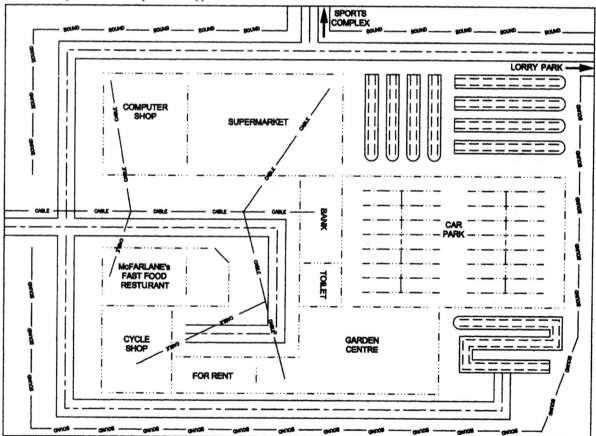

Figure 6.2 Shopping complex layout using created linetypes (simple, complex and multiline).

5   *Linetype appearance.*
The appearance of non-continuous lines is controlled by LTSCALE and is global, i.e. all linetypes are redrawn to suit the new value of LTSCALE. Release 13 introduced an option in the CHANGE and CHPROP commands – the ltScale option. This allows selected objects to have their linetype scale modified with respect to the current LTSCALE value. For instance, if the global value of LTSCALE is 2 and individual objects are selected then:
a) ltScale = 0.5, the effective LTSCALE value is 1
b) ltScale = 2, the effective LTSCALE value is 4

In Fig. 6.2, I had set LTSCALE to 2 and set the following ltScale option values:
a) NEWLINE1, ltScale = 0.2
b) NEWLINE2, ltScale = 0.3.

6   *Task.*
Optimize the LTSCALE value and the ltScale option values for your two new linetypes, then save the drawing as **R14CUST\SHOPLAY** as it will be used later in this chapter.

Complex linetypes containing text

Release 14 allows linetypes to be created which contain items of text, and several are available for use if loaded, e.g. GAS_LINE, HOT_WATER_SUPPLY, etc. Complex linetypes are:
a) created, loaded and set in the same manner as simple linetypes
b) have the same two line descriptor as simple linetypes
c) have the following descriptor format:
   *NAME, description
   A,pen movements,["TEXT",STYLE, R,A,S,*X*,*Y*]
d) it is the addition of the items in the [] brackets which allow text to be added to linetypes. These items are:
   i)   "TEXT"   the actual text item contained within ""
   ii) STYLE   a text style which must be loaded
   iii) R   the relative text rotation in the line
   iv) A   the absolute text rotation in the line
   v)   S   the scale factor for the text in the line
   vi) *X, Y*   the text offsets for the text item, computed from the end of the linetype definition vertex
e) the items "TEXT" and STYLE are mandatory but the rest are optional.

Creating/loading/setting/using two new complex linetypes

1   Still with the SHOPLAY drawing displayed?

2   Activate Notepad from the Windows taskbar and:
a) File–Open
b) alter file name to *.lin
c) pick the file r14cust\MYLINE
d) screen displays the two simple linetype descriptors
e) move cursor below the last line and add the following:
   *CABLE, a cable complex linetype
   A,12,–3,["CABLE",STDA3,R=0,S=1.1,*X*=–0.5,*Y*=–0.5],–6
   *BOUNDARY, __BOUND__BOUND__
   A,8,–2,["BOUND",STDA3,R=10,S=1.1,*X*=–0.25,*Y*=–0.75],–6
f) menu bar with File–Save to update MYLINE.lin
g) menu bar with File–Exit to return to AutoCAD (or minimize the Notepad screen for 'easy access' if needed).

3 Menu bar with Format–Layer and:
   a) make two new layers L3 and L4
   b) pick the L3 continuous linetype name to display the Select Linetype dialogue box
   c) pick Load to display the Load/Reload Linetypes dialogue box
   d) pick File
   e) scroll and pick r14cust\MYLINE then Open
   f) pick Cable then OK
   g) set linetype Cable to layer L3
   h) repeat the above procedure to set linetype Boundary to layer L4
   i) when the two new linetypes have been loaded and set to layers L3 and L4, pick OK from the layer control dialogue box.

4 Using layers L3 and L4, add boundary and cable lines to the layout.

5 Optimize the LTSCALE value and the ltScale option of the CHANGE command with your new linetypes.

6 When complete, save as R14CUST\SHOPLAY.

7 *Note.*
   To explain the complex linetypes, BOUNDARY will be used:
   a) *BOUNDARY, ___BOUND___BOUND___ : obvious?
   b) A,8,–2 : line for 8 drawing units and space for 2 drawing units
   c) [] : items to display text in the line
   d) "BOUND" : the item of text to be displayed
   e) STDA3 : the text style for the text item BOUND
   f) R=10 : the relative angle of the text in the line. Relative (R is recommended as it will rotate text as the drawn line is 'angled'. The absolute (A) entry does not allow this
   g) S=1.1 : the scale factor for the text item
   h) X=–0.25 : offset the text item to the left
   i) Y=–0.75 : offset the text item downwards to 'fit into the line'
   j) –6 : space after text item BOUND.

Multilines

Multilines are parallel lines which can consist of up to 16 **elements** and are created within AutoCAD by the user. A text editor is not used. The user defines the number of elements, the spacing between these elements as well as the element linetype, colour and end cap. Multilines have their own terminology as is displayed in Fig. 6.3.

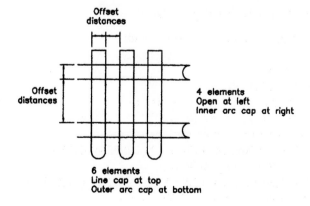

Figure 6.3 Basic multiline terminology.

Creating two new multilines

1 Still with the SHOPLAY drawing on the screen.

2 Menu bar with **Format–Multiline Style** and:
prompt   *Multiline Styles dialogue box*
with   1. STANDARD as current name
             2. graphical description (2 lines) in display area
respond   1. alter Name to: ML1
             2. enter Description: My first multiline
             3. pick Add – ML1 now current style
             4. pick Element Properties
prompt   *Element Properties dialogue box*
respond   1. pick the 0.5 BYLAYER BYLAYER line
              2. alter Offset to 1.5 then <R>
              3. pick −0.5 BYLAYER BYLAYER line
              4. alter Offset to 0 then <R>
              5. pick Add to create another 0 offset line
              6. pick this new 0 offset line
              7. alter Offset to −1.5 then <R>
              8. pick the 0.0 BYLAYER BYLAYER line
              9. pick Color, Red then OK
             10. pick Linetype, Center (loaded?) then OK
             11. Element Properties dialogue box as Fig. 6.4
             12. pick OK
prompt   *Multiline Styles dialogue box*
with   preview display of the three ML1 lines
respond   **pick Save**
prompt   *Save Multiline Styles dialogue box*
respond   1. alter Multiline name to: MYMLINE
             2. note file type extension: *.min
             3. pick Save
prompt   *Multiline Styles dialogue box*
respond   pick OK.

3 Repeat the menu bar selection Format–Multiline Style and:
prompt   *Multiline Styles dialogue box*
respond   1. scroll and pick STANDARD name
             2. alter Name to: ML2

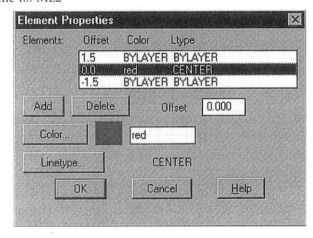

Figure 6.4 Element Properties dialogue box.

	3. enter Description: My second multiline
	4. pick Add
	5. pick Element Properties
prompt	*Element Properties dialogue box*
respond	1. pick 0.5 line and alter Offset to 1.5 then <R>
	2. pick −0.5 line and alter Offset to 0.75 then <R>
	3. pick Add to create another 0 offset line
	4. alter the offset of this new line to −0.75 then <R>
	5. pick Add to create another 0 offset line
	6. alter the offset of this new line to −1.5 then <R>
	7. alter the 0.75 and −0.75 lines with:
	a) colour: blue
	b) linetype: hidden (loaded?)
	8. pick OK
prompt	*Multiline Styles dialogue box*
with	preview display of the four elements
respond	**pick Multiline Properties**
prompt	*Multiline Properties dialogue box*
respond	set the following:
	a) Start cap: Line, i.e. tick in box
	b) End cap: Outer arc
	c) Angles: 90
	d) pick OK
prompt	*Multiline Styles dialogue box* as Fig. 6.5
respond	**pick Save**
prompt	*Save Multiline Styles dialogue box*
respond	1. ensure MYMLINE is current file name
	2. pick Save
prompt	*Multiline Styles dialogue box*
respond	pick OK.

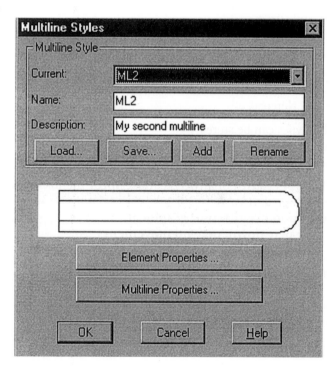

Figure 6.5 Multiline Styles dialogue box.

Using the two created multilines

1 Make a new layer ML, colour white and current.

2 Select the MULTILINE icon from the Draw toolbar and:
  prompt   *Justification/Scale/STyle/...*
  enter   **S <R>** – the scale option
  prompt   *Set Mline scale<1.00>* and enter: **4 <R>**
  prompt   *Justification/Scale/STyle/...*
  enter   **ST <R>** – the style option
  prompt   *Mstyle name* and enter: **ML1 <R>**
  prompt   *Justification/Scale/Style/<From point>*
  respond draw a 'road system' using the ML1 multiline style and right-click when
             required.

3 Menu bar with **Draw–Multiline** and:
  a) set the scale to 3
  b) set the style to ML2
  c) draw as required.

4 *Investigate*:
  a) the effect of CHANGE–ltScale on the multilines
  b) using the Change Properties icon
  c) optimize layout as Fig. 6.2 the save as R14CUST\SHOPLAY
  d) modifying intersection multilines with the menu bar sequence **Modify–Object–Multiline** – you should be able to 'figure out' the resulting dialogue box?

5 This exercise is now complete.

Summary

1 Linetypes can be:
  a) simple: dashes, dots and spaces
  b) complex: containing text items
  c) multi: consisting of different line elements.

2 Simple and complex linetypes can be created:
  a) from within AutoCAD with -LINETYPE
  b) using a text editor, e.g. Notepad.

3 Simple and complex linetypes can be stored:
  a) within the AutoCAD .lin file – not recommended
  b) in user-defined .lin files – recommended.

4 Simple and complex linetypes consist of a two line **descriptor** the format of which is very strict and cannot be altered. It is:
  *NAME, linetype description (written or graphical)
  A, coded definitions for pen–down, pen–up movements.

5 Complex linetypes contain field definitions within the coded part of the linetype descriptor. These field definitions determine how the text item will be aligned with the line. A complex linetype descriptor is:
  *NAME, description
  A, coded definitions, ["TEXT",ST,S,R,X,Y].

6 The procedure for creating new linetypes is:
   a) create in a .lin file
   b) load the linetypes from the .lin file
   c) set the linetypes in new layers
   d) use the new linetypes as required.

7 The appearance of linetypes is controlled by LTSCALE (globally) and the ltScale option of the Change Properties command.

8 Multilines are created within AutoCAD.

9 Multilines can consist of 16 elements with:
   a) different offsets
   b) varying colour and linetype.

10 Multilines are stored in user-defined .min files.

11 The appearance of multilines is controlled by the scale factor option, LTSCALE and ltScale.

Assignment

The shopping complex layout is adjacent to a sports complex due to the forward thinking of the local council. It is the athletics stadium of this sports complex which is the linetype activity.

Activity 6: Athletics stadium

1 Create three new linetypes using the information given below, the linetypes being:
   a) simple: for the stadium sectors, name: SECTOR
   b) complex: for the track outline, name: TRACK
   c) multiline: for the field, name: LANES.

2 Using the three created linetypes, design an athletic stadium of your choice.

3 Optimize the LTSCALE and ltScale values.

4 When complete, save as R14CUST\STADIUM for future recall.

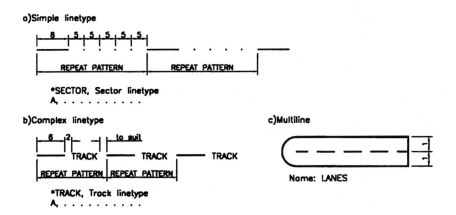

Chapter 7

Creating hatch patterns

Hatch pattern creation appears similar to linetype creation but is slightly more involved. Hatch patterns are stored in user-written files, and the extension is **.PAT**. The AutoCAD hatch patterns are stored within the **acad.pat** and **acadiso.pat** files in the SUPPORT sub-folder. New hatch patterns can be:
a) added to the AutoCAD .pat files
b) created in separate hatch pattern files.

In this chapter we will investigate one of the AutoCAD hatch patterns and then create several new hatch patterns in new user-defined files. We will not add to the existing AutoCAD hatch pattern files.

What are hatch patterns?

Hatch patterns consist of combinations of straight lines and spaces, i.e. pen down and pen up movements. The lines can be horizontal, vertical or inclined. Every hatch pattern must contain a minimum of two lines (most contain several lines), these lines being:
a) the header line
b) the pattern lines.

To view the AutoCAD R14 hatch patterns:

1 Open you R14CUST\STDA3 **DRAWING FILE** – not the template file.

2 Activate Notepad from the Windows taskbar and:
   prompt   *Untitled Notepad screen*
   respond   1. scroll and pick the AutoCAD R14\SUPPORT folder
              2. alter the File name to ***.pat** then <R>
              3. pick **acadiso** then Open.

3 The screen will display the AutoCAD Hatch Pattern file, one of these patterns being:
   *ANGLE, Angle steel
   0, 0,0, 0,6.985, 5.08,−1.905
   90, 0,0, 0,6.985, 5.08,−1.905.

4 Study some of the other patterns then exit Notepad.

Hatch pattern descriptors

Every hatch pattern consists of (at least) a two line **descriptor** made from three distinct parts:

1 The name of the hatch pattern preceded by an asterisk (*), e.g. *BRASS, *ANGLE. Capitals should be used for the name.

2 A written description of the hatch pattern. It follows the name and is separated from the name with a comma (,). The descriptive text is recommended as small letters.

3 The actual hatch pattern line(s) which are coded definitions for the lines which make up the hatch pattern. The definitions are:
   a) the ANGLE (ANG) from the horizontal
   b) the *X* ORIGIN (*X*) of the line
   c) the *Y* ORIGIN (*Y*) of the line
   d) the OFFSET DISPLACEMENT (*OX*) in the *X* direction
   e) the OFFSET DISPLACEMENT (*OY*) in the *Y* direction **perpendicular to the original pattern line**
   f) the pattern codes, i.e. pen down/up movements.

Notes on hatch pattern descriptors

1 The letters (*a*)–(*f*) will be referred to in our examples.

2 The hatch pattern name and descriptor are always written as the first line, e.g. *ANGLE, Angle steel. This line is called the **HEADER**.

3 The pen down and pen up movements are given as drawing units and one drawing unit is 1 mm.

4 There must be at least one pattern code line.

5 A maximum of six dashed lengths are permitted in any pattern line.

6 The format for the hatch pattern descriptor is very strict and cannot be altered. It is:
   *NAME, descriptive text
   ANG,*X*,*Y*,*OX*,*OY*, pen movements
   ANG,*X*,*Y*.*OX*,*OY*, pen movements, etc.

AutoCAD hatch pattern for discussion

This hatch pattern is
*ANGLE, Angle steel   – header line
0, 0,0, 0,6.985, 5.08,–1.905   – pattern line 1

The hatch pattern is detailed in Fig. 7.1 with:
a) the hatch pattern descriptor
b) the basic hatch 'element' with sizes as drawing units
c) the hatch pattern construction with various sizes as drawing units
d) using the hatch pattern at varying scales and angles.

The following notes are relevant to this hatch pattern:

1 Header line: name of pattern   *ANGLE
                  description   Angle steel

2 Pattern line 1:   0,0,0,0,6.985,5.08,–1.905
   a) 0 angle, i.e. a horizontal line – the original pattern line 1
   b) 0 *X* origin point
   c) 0 *Y* origin point, i.e. pattern line 1 starts at an assumed (0,0) point
   d) 0 *X* offset displacement, i.e. there is no horizontal displacement in the *X* direction between successive lines of this type
   e) 6.985 *Y* offset displacement, i.e. there is a 6.985 drawing unit vertical displacement **perpendicular** to the original pattern line 1 between successive lines of this type
   f) 5.08 and –1.905 are the pen movements, i.e. pen down for 5.08 drawing units the pen up for 1.905 drawing units. This will be repeated until the hatch boundary is reached.

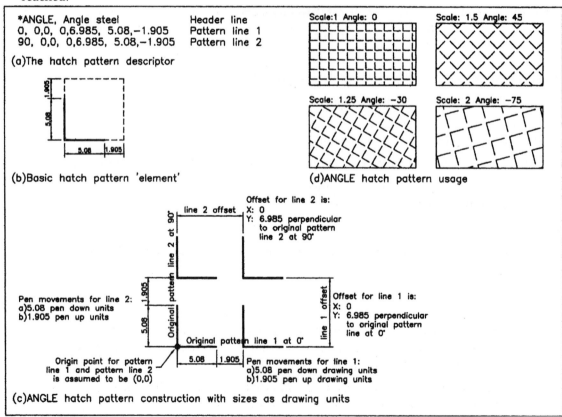

Figure 7.1 Details of the acadiso.pat ANGLE hatch pattern – description, construction and usage.

3 Pattern line 2:   90,0,0,0,6.958,5.08,−1.905
  a) 90 angle, i.e. a vertical line – the original pattern line 2
  b) 0 X origin point
  c) 0 Y origin point, i.e. the assumed origin point for pattern line 2 is (0,0) which is the same origin point as for pattern line 1. Thus pattern lines 1 and 2 start at the same point
  d) 0 X offset displacement, i.e. there is no horizontal displacement in the X direction between successive lines of this type
  e) 6.958 Y offset displacement, i.e. there is a 6.985 drawing unit vertical displacement **perpendicular** to the original pattern line 2 between successive lines of this type
  f) 5.08 and −1.905 are the pen movements, i.e. pen down for 5.08 drawing units and pen up for 1.905 drawing units. This will be repeated until the hatch boundary is reached.

Note

1 Probably the most difficult and confusing concept in hatch pattern design is the X and Y offset displacements. This note will (hopefully) remove any confusion.

2 a) the X offset displacement is **always in the direction of the original pattern line**
  b) the Y offset displacement is **always perpendicular to the direction of the original pattern line**.

3 Thus if the original pattern line is vertical:
  a) the X offset displacement is in the traditional y direction
  b) the Y offset displacement is in the traditional x direction.

4 Refer to pattern lines 1 and 2 in Fig. 7.1 and you will realize that the X and Y offset displacement concept is relatively straightforward. The phrase **perpendicular to the original pattern line** is important.

Designing our own hatch patterns

As stated earlier, new hatch patterns can be created:
a) within the existing AutoCAD .pat files
b) in user-defined .pat files.

We will create our own hatch pattern files in our R14CUST folder and the following points are **very important**:

1 Only one pattern can be stored in the file.

2 The file **must** have the same name as the hatch pattern.

3 The hatch pattern should be stored in the same folder as the drawing in which it is to be used – R14CUST.

Note

1 When I design hatch patterns I try to work with a basic shape on square paper. This basic shape consists of one complete element of the paper. I then increase this basic shape to four shapes in a rectangular type pattern.

2 Ensure that all the hatch pattern exercises are completed and saved, as they will be used in a later chapter for hatch pattern palette (icon) creation.

Hatch pattern design 1 – hang shapes

1 STDA3 **DRAWING** file on screen. Refer to Fig. 7.2 which displays the following information about the pattern to be created:
   a) the basic element with sizes as drawing units
   b) details of line identification, line origins, line offset and the line pen movements
   c) the hatch pattern descriptor for HANG
   d) using the created hatch pattern.

2 Using the windows taskbar open Notepad and:
   prompt   untitled Notepad screen
   respond   *a*) enter the following lines:
   　　　　　 *HANG, a 3 line pattern<R>
   　　　　　 90,0,0,0,10,6,–4<R>
   　　　　　 90,5,4,0,10,2,–8<R>
   　　　　　 0,0,6,0,10,5,–5<R>
   　　　 b) menu bar with File–Save As and:
   　　　　　 1. ensure r14cust is current
   　　　　　 2. enter File name: **HANG.pat**
   　　　　　 3. pick Save
   　　　 c) menu bar with File–Exit to return to AutoCAD.

3 With later OUT current, draw some rectangular shapes.

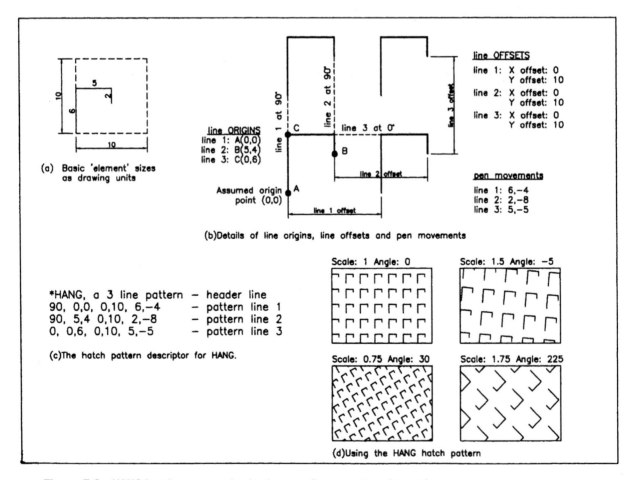

Figure 7.2   HANG hatch pattern – basic element sizes, construction and usage.

4 Make layer SECT current, enter **HATCH <R>** at the command line and:
   prompt   *Enter pattern name* and enter: **HANG <R>**
   prompt   *Scale for pattern* and enter: **1 <R>**
   prompt   *Angle for pattern* and enter: **0 <R>**
   prompt   *Select hatch boundaries*
   respond **window one of the rectangles then right-click**.

5 If the HANG hatch pattern has been created correctly, hatching will be added to the selected rectangle.

6 Select the HATCH icon from the Draw toolbar and:
   prompt   *Boundary Hatch dialogue box*
   with   No icon and Custom Pattern: HANG
   respond 1. alter Scale to: 1.5
              2. alter Angle to: −5
              3. pick Pick Points< and pick any internal point in another rectangle then right-click
   prompt   *Boundary Hatch dialogue box*
   respond Preview, Continue, Apply.

7 Add the HANG hatching to the other rectangles, altering the scale and angle values.

8 Save the layout if required.

9 *Hatch errors.*
   Errors do occur when creating hatch patterns, two of these being:
   a) Bad pattern definition error. This is usually a result of a period (.) being used instead of a comma (,). A message is normally displayed indicated the line in the hatch pattern file. The file has to be altered.
   b) The hatching seems to take some time to fill the shape. This usually means that the line descriptors have been wrongly written and must be altered. This error is more common when using the command line HATCH. The dialogue box Preview option generally allows the user to 'see' if the added hatching is correct.

10 *Note*: the appearance of hatch patterns is controlled by the scale factor. LTSCALE has no effect on hatch patterns.

Hatch pattern design 2 – double squares

This pattern is slightly more complex than the first exercise. It consists of two squares and requires eight lines to be defined.

1 Open drawing file R14CUST\STDA3 and refer to Fig. 7.3 which displays:
   a) the basic element with sizes as drawing units
   b) four repeating elements in the pattern with line identification
   c) the eight line origins
   d) the information required to make the eight line descriptors
   e) different pattern usage.

2 Activate Notepad from the Windows taskbar and enter the following lines, remembering to press <R> at the end of each line:
   *DBOX, a double box pattern
   0,0,0,0,8,6,–2
   0,2,2,0,8,2,–6
   0,2,4,0,8,2,–6
   0,0,6,0,8,6,–2
   90,0,0,0,8,6,–2
   90,2,2,0,8,2,–6
   90,4,2,0,8,2,–6
   90,6,0,0,8,6,–2.

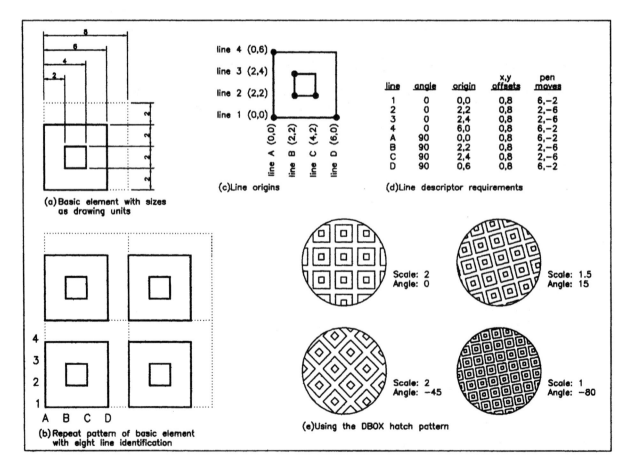

Figure 7.3 DBOX hatch pattern – basic element, line identification, descriptors and usage.

3 Menu bar with **File–Save As** and:
   a) r14cust folder current
   b) enter file name: **DBOX.pat**
   c) pick Save
   d) menu bar with File–Exit to return to AutoCAD.

4 With layer OUT current draw some circles.

5 Make layer SECT current and select the HATCH icon from the Draw toolbar and:
   prompt   *Boundary Hatch dialogue box*
   respond   1. Pattern type: pick **Custom**
               2. Custom pattern: enter **DBOX**
               3. Scale: enter **2**
               4. Angle: enter **0**
               5. pick **Pick Points<**
               6. pick any point within a circle then right-click
               7. preview–continue–apply (if hatching is 'correct').

6 Hatch the other circles using different scale and angle values.

7 *Investigate*.
   We have used our created hatch patterns (HANG and DBOX) with the STDA3 **DRAWING** file. Will it work with the STDA3 template file? I found that I could not activate the hatch patterns when I had opened the template file, even the STDA3 template file from the r14cust folder. The message displayed was 'Unknown pattern name'.

8 This exercise can now be saved if required.

Hatch pattern design 3 – a weave pattern

This hatch pattern also consists of eight lines, but the effect is entirely different from the eight line double box pattern. Figure 7.4 details the pattern with:
a) the basic element with sizes as drawing units
b) four repeating elements in the pattern with line identification
c) the eight line origins
d) the information required for the line descriptors
e) using the weave pattern.

The procedure is as before:

1 Open your STDA3 **DRAWING** file.

2 Enter Notepad and enter the following lines (remember <R>):
   *WEAVE, a weave pattern effect
   0,0,0,0,6,4.5,–1.5
   0,0,1.5,0,6,1.5,–1.5,3
   0,0,3,0,6,1.5,–1.5,3
   0,0,4.5,0,6,4.5,–1.5
   90,0,0,0,6,1.5,–1.5,3
   90,1.5,0,0,6,4.5,–1.5
   90,3,0,0,6,4.5,–1.5
   90,4.5,0,0,6,1.5,–1.5,3.

3 *a*) File–Save as with name **r14cust\WEAVE.pat**
   b) File–Exit to return to AutoCAD.

4 With layer OUT current draw some shapes.

5 With layer SECT current, hatch the shapes using the WEAVE pattern and alter the scale and angle values.

6 Save if required.

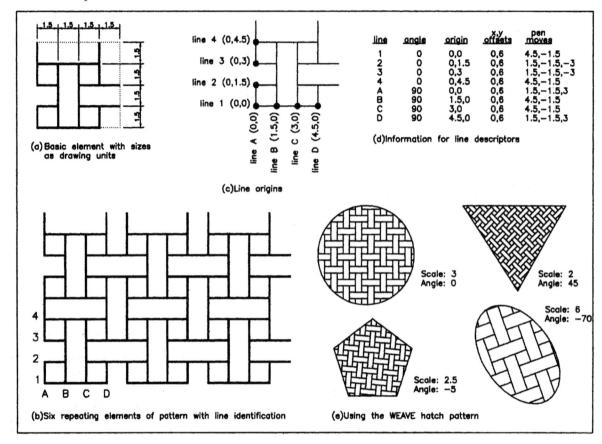

Figure 7.4 WEAVE hatch pattern – basic element, construction and usage.

Hatch pattern design 4 – an offset L pattern

This pattern uses X and Y offsets to give an interesting effect. Refer to Fig. 7.5 which displays:

a) the basic element with drawing units sizes

b) repeating patterns with the offsets

c) repeating patterns with the pen movements

d) information for the line descriptors

e) using the pattern.

Same procedure as before, i.e.

1 Open the STDA3 drawing file.

2 Activate Notepad and enter the following lines:
   *ELLS, an offset L shaped pattern
   0,0,0,10,10,15,–15
   0,5,5,10,10,10,–20
   0,0,15,10,10,5,–25
   90,0,0,–10,10,15,–15
   90,5,5,–10,10,10,–20
   90,15,0,–10,10,5,–25.

3 Menu bar and:
   a) File–Save as with the name r14cust\ELLS.pat then Save
   b) File–Exit to return to AutoCAD.

4 Add the ELLS hatching to some shapes.

5 Save the layout if required.

6 *Note*.
   Probably the hardest part of this hatch pattern to understand is the *X* and *Y* offset. Figure 7.5(b) gives these offsets for lines 1–3, those for lines 4–6 being similar. For lines 1–3 both the *X* and *Y* offsets are 10. For lines 4–6 the offsets are slightly different. Can you reason out why?

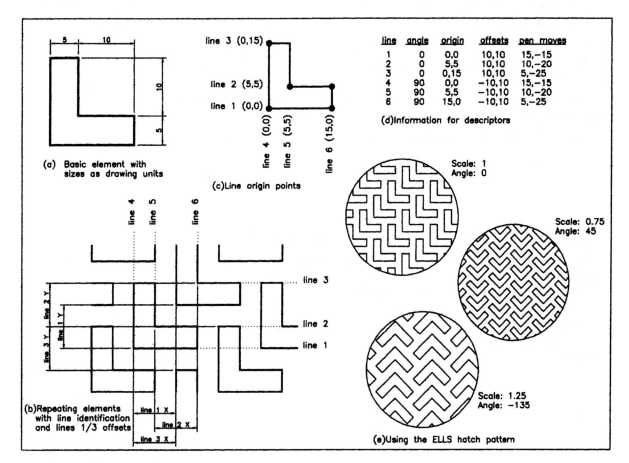

Figure 7.5 The ELLS hatch pattern.

Hatch pattern design 5 – an arrow pattern

This pattern only consists of four lines but is quite complex as it contains 'angled' lines. When hatch patterns have lines which are not horizontal or vertical, the offsets usually require the use of the trigonometric functions to obtain the actual offset distance values. My basic arrow shape consisted of the traditional 3, 4, 5 right-angled triangle, selected to make the pen up/down movements easy. The angles associated with these triangle side values are and 143.1301.

1 Open your STDA3 drawing file and refer to Fig. 7.6 which gives details of:
   a) the basic arrow shape with drawing unit sizes
   b) a repeat pattern with line identification and offset distances
   c) the line origin points
   d) information for the line descriptors
   e) using the hatch pattern
   f) *note*: if you are unsure of the values in the line descriptor, draw the arrow to scale, multiple copy it and measure the various distances.

2 Activate Notepad and enter the following lines:
   *ARROWS, an open arrow hatch pattern
   0,0,5,16,12,8,–24
   36.8699,0,5,14.4,19.2,5,–15
   90,4,0,12,16,8,–16
   143.1301,8,5,5.6,19.2,5,–15.

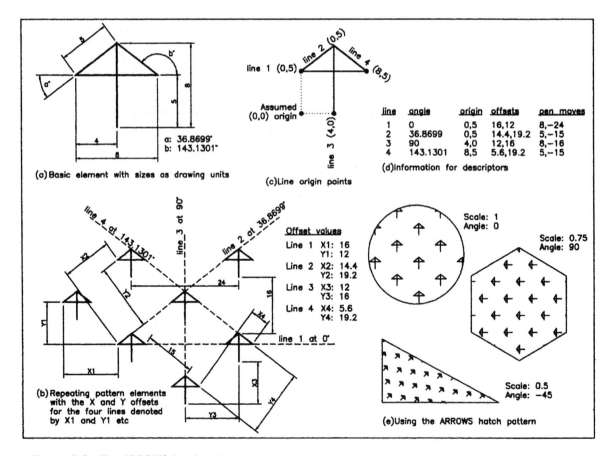

Figure 7.6 The ARROWS hatch pattern.

3 Menu bar with:
   a) File–Save As with the name: **r14cust\ARROWS.pat**
   b) File–Exit to return to AutoCAD.

4 Draw some shapes and add the ARROWS hatching – scales and angles to suit.

5 Save the drawing if required.
   Note: this pattern is produced by permission of Bob McNay, one of my CAD lecturers.

Task

Open the drawing R14cust/WINDOW to display the original attribute layout. Refer to
Fig. 7.7 and using this layout:
a) add the five created hatch patterns to the computer monitors using your own scale
    and angle values
b) you have to remove the attribute text – easy?
c) when complete, save the drawing although we will not use it again.

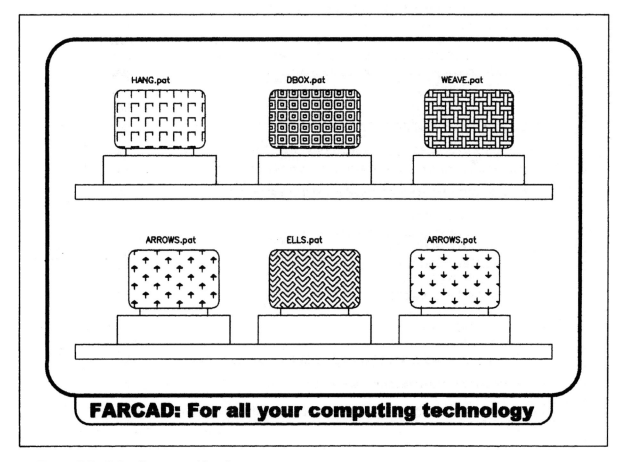

Figure 7.7 Using five created hatch patterns.

Summary

1 Hatch patterns consist of a series of parallel lines, these lines being horizontal, vertical or inclined.

2 The format for creating hatch patterns is very rigid and cannot be altered. The hatch pattern **descriptor** is:
*NAME, description
angle, *X–Y* origin, *X–Y* offsets, pen movements.

3 The first line in the hatch pattern descriptor is called the header line. The other lines are called pattern definition lines.

4 Created hatch patterns can be created:
a) within the AutoCAD acad.pat and acadiso.pat files
b) in new .pat files.

5 It is strongly recommended that new hatch patterns are created in new files.

6 Only one hatch pattern is permitted in any new file and **the pattern name must be the same as the file name**.

7 The *X* and *Y* offset displacements are probably the most difficult concept to understand with defining hatch patterns (especially the *Y* offset)
a) the *X* offset is always in the direction of the original pattern line
b) the *Y* offset is always PERPENDICULAR to the direction of the original pattern line.

8 Hatch patterns are not affected by the LTSCALE system variable. It is the hatch pattern scale factor that determines the hatch appearance.

9 Customized hatch patterns can be activated from the command line, by icon selection or from the menu bar. If the hatch dialogue box is being used, there will be no icons displayed for the created patterns, as these have not yet been 'programmed'. This will be investigated in a later chapter.

Assignment

The five created hatch patterns have been listed in full, and it has simply been a matter of entering the lines using Notepad. To test your understanding of how hatch patterns are customized, I have included an activity which requires two new patterns to be created.

Activity 7: Lorry park

1 Open the lorry park layout from Activity 1(b).

2 Erase the attribute data.

3 Create the two given hatch patterns.

4 Add hatching to the lorries with your own scale and angle values.

Four additional hatch patterns for consideration

Before leaving this chapter on hatch pattern creation/design, I have included another four patterns for your consideration, shown in Fig. 7.8. These patterns were designed by Charles Sweeney, a student on one of my HNC Computer Aided Draughting and Design (CADD) courses. Students on this course have to complete a CADD Project, the topic being of their own choosing. When I was teaching hatch pattern design, I made the comment that 'there were no circular type hatch patterns'. Charles decided to investigate this as a possible project and designed about 50 new hatch patterns, the four included being from his project portfolio. He concluded that while there are no circular hatch patterns, a curved '**effect**' could be obtained. Two of these hatch patterns give this effect.

Further details of the other hatch patterns can be obtained by contacting me directly.

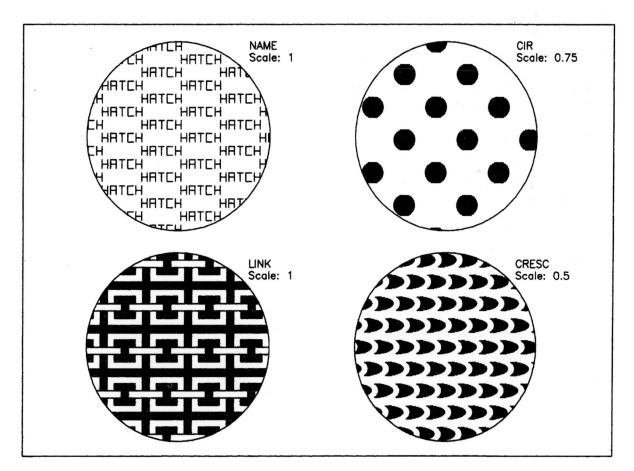

Figure 7.8   Four additional student-created hatch patterns (courtesy of Charles Sweeney).

First additional hatch pattern

*NAME, hatch word pattern
90,0,0,9,28,6,–12
0,0,3,28,9,4,–52
90,4,0,9,28,6,–12
90,6,0,9,28,6,–12
0,6,6,28,9,4,–52
0,6,3,28,9,4,–52
90,10,0,9,28,6,–12
90,14,0,9,28,6,–12
0,12,6,28,9,4,–52
90,18,0,9,28,6,–12
0,18,6,28,9,4,–52
0,18,0,28,9,4,–52
90,24,0,9,28,6,–12
0,24,3,28,9,4,–52
90,28,0,9,28,6,–12

Second additional hatch pattern

*CIR, filled circle effect
0,55.64,41,30,30,8.72,–51.28
0,54,42,30,30,12,–48
0,52.86,43,30,30,14.28,–45.72
0,52,44,30,30,16,–44
0,51.34,45,30,30,17.32,–42.68
0,50.83,46,30,30,18.33,–41.67
0,50.46,47,30,30,19.08,–40.92
0,50.20,48,30,30,19.6,–40.4
0,50.05,49,30,30,19.9,–40.1
0,50,50,30,30,20,–40
0,55.64,59,30,30,8.72,–51.28
0,54,58,30,30,12,–48
0,52.86,57,30,30,14.28,–45.72
0,52,56,30,30,16,–44
0,51.34,55,30,30,17.32,–42.68
0,50.83,54,30,30,18.33,–41.67
0,50.46,53,30,30,19.08,–40.92
0,50.20,52,30,30,19.6,–40.4
0,50.05,51,30,30,19.9,–40.1

Third additional hatch pattern

*CRESC, relief crescent shape
0,10,0,15,30,20,–10
0,9.949,1,15,30,19.898,–10.102
0,9.797,2,15,30,19.594,–10.406
0,9.539,3,15,30,19.078,–10.922
0,9.165,4,15,30,18.33,–11.67
0,8.660,5,15,30,17.32,–12.68
0,8,6,15,30,16,–14

0,7.141,7,15,30,14.282,–15.718
0,6,8,15,30,12,–18
0,4.358,9,15,30,8.716,–21.284
0,9.949,–1,15,30,19.898,–10.102
0,9.797,–2,15,30,19.594,–10.406
0,9.539,–3,15,30,19.078,–10.922
0,9.165,–4,15,30,18.33,–11.67
0,8.660,–5,15,30,17.32,–12.68
0,8,–6,15,30,16,–14
0,7.141,–7,15,30,14.282,–15.718
0,6,–8,15,30,12,–18
0,4.358,–9.15,15,30,8.716,–21.284

Fourth additional hatch pattern

*LINK, chain link with shaded background
90,0,0,0,30,10,–5
90,5,5,0,15,5,–5,5,–15
90,10,10,0,30,5,–25
90,15,10,0,30,5,–25
90,25,0,0,30,10,–5
0,0,0,0,30,25
0,5,5,0,15,15,–5,5,–5
0,15,10,0,30,25
0,15,15,0,30,25
0,0,25,0,30,25
0,5,16,0,30,15,–5,5,–5
0,5,17,0,30,15,–5,5,–5
0,5,18,0,30,15,–5,5,–5
0,5,19,0,30,15,–5,5,–5
0,5,6,0,30,15,–5,5,–5
0,5,7,0,30,15,–5,5,–5
0,5,8,0,30,15,–5,5,–5
0,5,9,0,30,15,–5,5,–5
0,10,11,0,30,5,–25
0,10,12,0,30,5,–25
0,10,12,0,30,5,–25
0,10,13,0,30,5,–25
0,10,14,0,30,5,–25
0,0,26,0,30
0,0,27,0,30
0,0,28,0,30
0,0,29,0,30
0,25,1,0,30,5,–25
0,25,2,0,30,5,–25
0,25,3,0,30,5,–25
0,25,4,0,30,5,–25
0,25,21,0,30,5,–25
0,25,22,0,30,5,–25
0,25,23,0,30,5,–25
0,25,24,0,30,5,–25

Chapter 8

Creating shapes

When a drawing object has to be used several times, the simplest way of achieving this is usually by creating and inserting a block. It may be necessary to insert the block repeatedly, e.g. when drawing an electrical/pneumatic circuit and the short delay experienced when inserting the block becomes noticeable and can slow down the draughting process. Shapes help overcome this problem.

What are shapes?

Shapes appear similar to blocks. They are drawing objects stored in the same way that AutoCAD stores its text fonts. The method of defining a shape is very concise and efficient, but it is not as easy to create a shape as it is to make a block.

Despite any similarity in use, shapes and blocks have basic differences in the way they are defined. A block is a collection of drawn objects that are 'joined together' to form a single object. They are made with using straightforward drawing techniques and are saved either as part of the drawing file in which they were created (BLOCKS) or as separate drawing files (WBLOCKS) which are available for use by other operators. Whole drawings can be inserted into other drawings as wblocks. The use of blocks saves memory in comparison to the multiple copying of items. Blocks also have the added capacity of attributes, making the text information contained in a block available for extraction for use in other software packages.

Shapes are made from lines and arcs and are very suitable for items such as electrical symbols, alphabets, etc. Shapes are 'stored' in their own files which have the extension **.SHP** and are written using a text editor, e.g. Notepad. The shape files can be stored in named folders or on floppy disk. The AutoCAD text font files are created from shapes.

The process of defining shapes requires the very formal specification of a number of parameters. Shape definitions appear to be similar to linetype and hatch pattern descriptors, but whereas linetypes and hatch patterns are written in a format suitable to a plotter (i.e. pen down and up movements), shapes are written as **screen vectors** in hexadecimal code.

Note

1 Screen vectors: are instructions stored in memory to position lines and circles on the screen.

2 Hexadecimal: is a number system with a base of 16. It is a standard computer numbering system and:
   Base 10:   0 1 2 3 4 5 6 7 8 9 10 11 12 13 14 15
   Hexadecimal equivalent: 0 1 2 3 4 5 6 7 8 9 A B C D E F

3 Computers use hexadecimal numbers from '00' to 'FF', i.e. 0 to 255 and the memory space which stores these pairs of digits (the 00 to FF) is called a **byte** and each byte is divided into two **nibbles**.

Shape descriptors

1 Every shape in a file **must** have its own unique two line definition (descriptor) and the format is:
*shapenumber, number of bytes, SHAPENAME
coded description.

2 The shapenumber is between 1 and 255 in any one file and **must be preceded by an asterisk (*)**, e.g. *4, *14, *144.

3 Number of bytes: this is the total number of bytes in the actual shape definition. It is obtained by adding the bytes in the coded description line. A maximum of 2000 bytes is permitted per shape.

4 SHAPENAME: is the name used to recall the shape when it is to be used in a drawing. It must be in **CAPITAL LETTERS**, e.g. ARROW, RES.

5 Coded description: is a series of codes which relate to the vector direction and length, as well as other specialist codes.

6 A typical shape definition could be:
*18,7,LSHP
050,02C,020,03C,078,054,0
a) 18: the shape number in the file, preceded by *
b) 7: the number of bytes in the description line
c) LSHP: the shape name in capitals
d) 050,02C, etc.; the coded descriptions of the vectors which make up the shape.

7 *Note.*
a) the coded description line is usually written first to obtain the number of bytes which make up the shape
b) all hexadecimal letters in the coded description must be in CAPITALS
c) every line descriptor ends with a 0.
In this chapter we will investigate how shapes are created with a series of different exercises. The shapes which will be considered are made from:
a) predefined vectors
b) variable X and Y vectors
c) pen down and up movements
d) arc vectors.

Note

When creating shapes, the following procedure is adopted:

1 Draw the shape to be created.

2 Code the shape definition.

3 Write and save the shape file using a text editor.

4 Compile the shape file in AutoCAD.

5 Load the shape file.

6 Insert the shapes into the drawing with the SHAPE command.

Predefined vectors

There are 16 predefined vector directions which can be used to create straight line elements in a shape. As an example of how predefined vectors are used, a vertical line drawn from bottom to top of length 5 drawing units would be coded as **054**. If the line were drawn from top to bottom the code would be **05C**. Predefined vectors are always defined in a **three digit format** as follows:

0LD   0: the leading 0 indicating a hexadecimal number to follow
           L: the **length** of the vector
           D: the **predefined direction** of the vector.

1 Open the STDA3 template file and refer to Fig. 8.1 which displays the 16 predefined vector directions – Fig. (a). This sketch should be used as a reference.

2 Figure 8.1(b) displays:
    a) the three shapes to be created with sizes as drawing units
    b) how the shapes have been 'drawn'
    c) the shape definitions.

3 Activate Notepad from the Windows taskbar and enter the following lines of text:
    *18,7,LSHP
    050,02C,020,03C,078,054,0
    *19,7,CHEV
    03E,032,06C,036,03A,064,0
    *20,5,PARM
    06F,04D,067,045,0

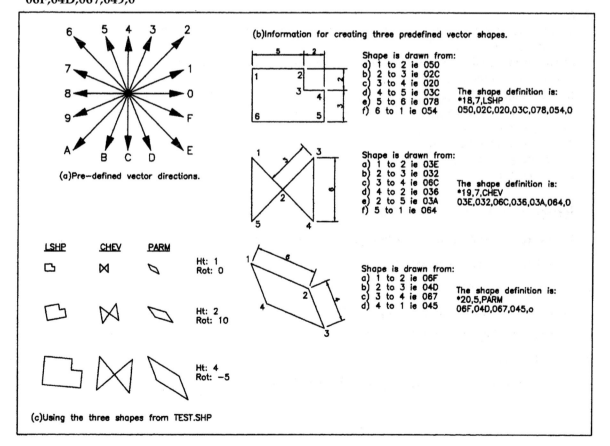

Figure 8.1 Predefined vector shapes for TEST.SHP.

4 Menu bar with File–Save As and:
   a) ensure r14cust is current
   b) enter file name as **TEST.SHP**
   c) pick Save
   d) menu bar with File–Exit to return to AutoCAD.

5 At the command line enter **COMPILE <R>** and:
   prompt   *Select Shape or Font File dialogue box*
   respond 1. Look in r14cust folder
              2. pick **Test**
              3. pick Open
   prompt   *Compiling shape/font description file*
              Compilation successful
              Output file C:\r14cust\Test.shx contains 79 bytes.

6 At the command line enter **LOAD <R>** and:
   prompt   *Select Shape File dialogue box*
   respond 1. Look in r14cust folder – note *.shx file type
              2. pick **Test** then Open
   and   nothing appears to have happened.

7 At the command line enter **SHAPE <R>** and:
   prompt   *Shape name (or ?)* and enter: **LSHP <R>**
   prompt   *Height* and enter: **1 <R>**
   prompt   *Rotation angle* and enter: **0 <R>**
   and   the LSHP will be displayed at selected point.

8 Repeat the SHAPE command, entering the shape names CHEV and PARM both with height: 1 and rotation angle: 0.

9 Insert the three shapes at other heights and angles – Fig. (c).

10 *Note.*
   The steps described above are always necessary when creating shapes, i.e.
   1. write and save the .SHP text file using Notepad.
   2. COMPILE the .SHP file into a .SHX file.
   3. LOAD the compiled .SHX file.
   4. use the SHAPE command to insert the shapes.

Variable X and Y vectors

The predefined vectors are very useful but not all shapes can be created with them. When a vector is required which does not conform to the predefined directions, vector control codes need to be used. There are two control codes governing variable X–Y vector movement, these being:
a) code 008: allows a single X–Y vector movement. This code can be enter as 8
b) code 009: allows multiple X–Y vector movement and must be ended with 0,0. This code can be entered as 9.

1 Open your STDA3 template file and refer to Fig. 8.2 which displays the five shapes to be created with:
   a) sizes as drawing units
   b) how the shapes are to be 'drawn'
   c) the shape definitions
   d) inserting the shapes into the drawing.

2 Activate Notepad from the Windows taskbar and enter the following lines:
*59,6,CLIP
044,8,2,–3,018,0
*60,6,STEP
030,8,2,3,040,0
*61,11,TAB
058,044,050,9,3,–2,–3,–2,0,0,0
*62,11,TEMP
054,8,3,1,02C,010,01C,8,–4,–3,0
*63,16,SAW
9,2,3,0,–3,2,3,0,–3,2,3,0,–3,0,0,0

3 Save the file as: r14cust\MYSHP1.SHP then return to AutoCAD.

4 At the command line enter:
   a) COMPILE <R> and open the file r14cust\Myshp1 – an Shp file
   b) LOAD <R> and open the file r14cust\Myshp1 – an Shx file.

5 At the command line enter SHAPE <R> and enter each of the five shape names (CLIP, STEP, TAB, TEMP, SAW) with:
   a) selected insertion points
   b) at varying heights and rotation angles.

6 Save if required.

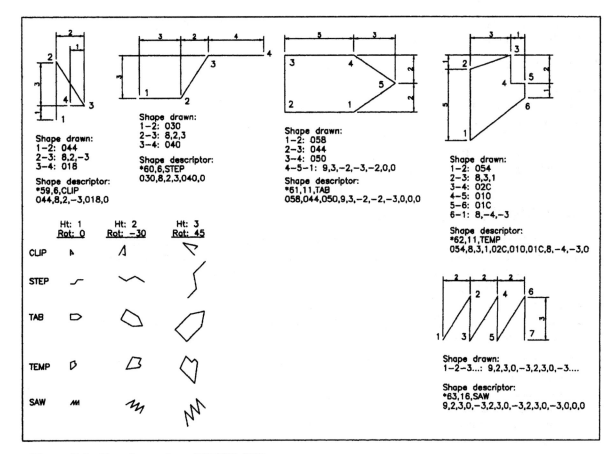

Figure 8.2 Five shapes from MYSHP1.SHP.

7 *Note.*

a) When writing the shape file in Notepad, brackets can be added around the *X–Y* coordinates, e.g. 8,2,–3 could be entered as 8,(2,–3). This may help the user 'identify' the entered vector values. The brackets are not counted as bytes

b) all shape definitions could be entered using the 9 code, e.g. the shape definition for TEMP could have been written as:

*62,16,TEMP
9,(0,5),(3,1),(0,–2),(1,0),(0,–1),(–4,–3),(0,0),0

c) the shape insertion point is at the 'definition start point'

d) the shape numbers need not be consecutive in the shape file.

Pen down and pen up codes

When shapes are being created, it is possible to have the pen down (draw mode) or up (non-draw mode). The pen modes are controlled by two control codes:

a) code 001: pen down – the default mode and need not be defined. The code can be entered as 1

b) code 002: pen up mode, and can be entered as 2.

These codes will be demonstrated with three new shapes written in a new .SHP file.

1 Open the STDA3 template file and refer to Fig. 8.3 which displays:

a) the three shapes to be created with sizes as drawing units

b) the shape definitions

c) inserting the compiled and loaded shapes into the drawing.

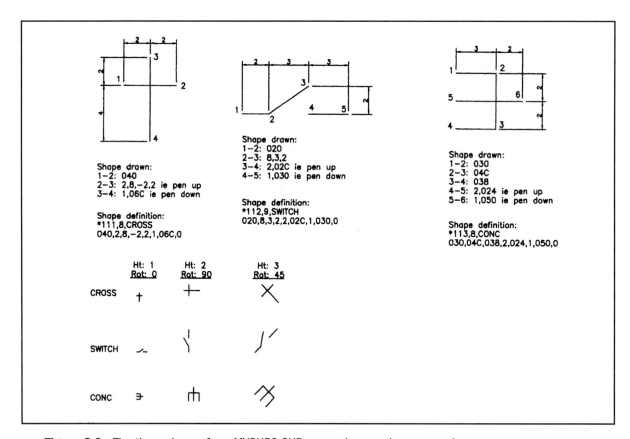

Figure 8.3 The three shapes from MYSHP2.SHP – pen down and pen up codes.

2   Activate Notepad and enter the following lines:
    *111,8,CROSS
    040,2,8,–2,2,1,06C,0
    *112,9,SWITCH
    020,8,3,2,2,02C,1,030,0
    *113,8,CONC
    030,04C,038,2,024,1,050,0

3   Save this file as r14cust\MYSHP2.SHP then return to AutoCAD.

4   *a*) Compile the r14cust\MYSHP2.SHP file.
    b) Load the compiled r14cust\MYSHP2.shx file.
    c) Insert the three shapes CROSS, SWITCH and CONC at different points with varying height and rotation.
    d) Save if required.

5   *Note.*
    The definition for the CROSS shape uses three control codes:
    2: pen up
    8: single *X–Y* vector movement
    1: pen down.

Shape file errors

When a shape file is being compiled, errors may be detected in the user's written file. AutoCAD will display the message:

'Bad shape definition at line? of ???'
'Error message'

Some of the error messages are:
a) premature end of file: missing 0 at end of second line
b) shape exceeds specific length: number of bytes in line 1 does not agree with the number of bytes in line 2
c) invalid shape element or bad syntax: using a (.) instead of a (,)
d) other errors include:
   – not starting with a *
   – using o instead of 0
   – not using CAPITALS for the name
   – not using a shape number between 1 and 255.

If errors are detected in the user's shape file they must be corrected by rewriting the shape file. The compile and load commands must then be used again but:
a) either a new template file must be opened
b) or AutoCAD must be exited then restarted.

A modified shape file **cannot be recompiled** in the existing drawing.

Curved vectors

Shape curved vectors are drawn as **octantarcs** (one-eighth of a circle) and require the control code 10 to be used. Curved vectors have their own format which is:

10,RAD,sign0ST with

10:	octant arc code
RAD:	arc radius
sign:	+ if drawn anti-clockwise
	– if drawn clockwise
0:	leading zero (needed)
S:	octant arc start number
T:	number of octant arcs 'moved through'

The curved vectors will be demonstrated with six new shapes in another new shape file, so:

1 Open the STDA3 template file and refer to Fig. 8.4 which displays:
   a) the octant arc number diagram – Fig. (a)
   b) two octant arc shape definitions as examples – Fig. (b)
   c) the six shapes to be created
   d) using the six new shapes.

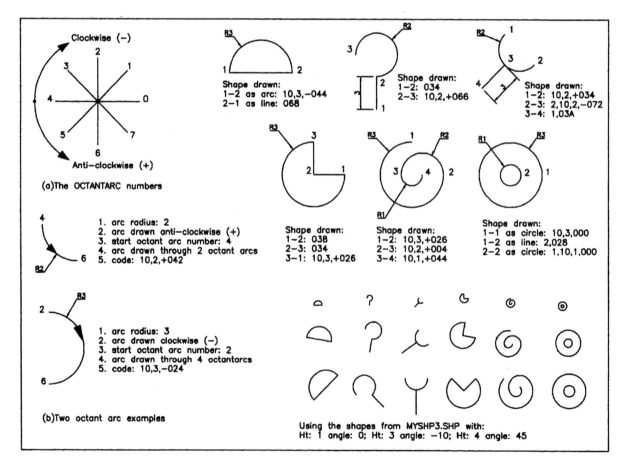

Figure 8.4   Six shapes from MYSHP3.SHP – curved vectors.

2 Activate Notepad and enter the following lines:
   *200,5,SEMI
   10,3,–044,068,0
   *201,5,HOOK
   034,10,2,+066,0
   *202,10,SAT
   10,2,+034,2,10,2,–072,1,03A,0
   *203,6,PAC
   038,034,10,3,+026,0
   *204,10,SPIRAL
   10,3,+026,10,2,+004,10,1,+044,0
   *205,10,CIRCS
   10,3,000,2,028,1,10,1,000,0

3 When written, save the file as r14cust\MYSHP3.SHP then return to AutoCAD.

4 *a*) Compile the MYSHP3.SHP file
   b) load the compiled MYSHP3.SHX file
   c) insert the six new shapes at varying heights and rotation angles
   d) save if required.

5 *Note.*
   a) The shape SAT uses the pen up/down codes (2 and 1) to move 'back along' the arc before drawing the line.
   b) The CIRCS shapes creates complete circles with the curved vector code 10,RAD,000 with:
      0: leading 0
      0: start octant arc number
      0: entering 0 as the number of octantarcs moved through gives a complete circle.
   c) The CIRCS shape also uses the pen up/down codes to 'move' between the two circles.

6 *Task.*
   a) Four .SHP shape files have been written. Using the drawing from the attribute exercise (r14cust\WINDOW), compile, load and use all the created shapes in the one drawing. Refer to Fig. 8.5 on the next page for a typical layout.
   b) Investigate the ? option of the SHAPE command.

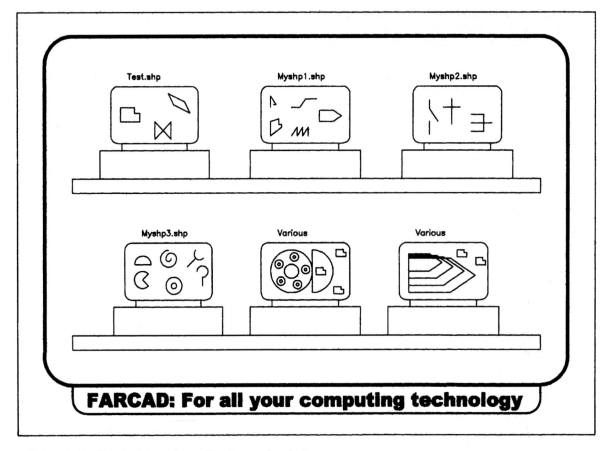

Figure 8.5 Using all four shape files in one drawing.

Complex linetypes containing shapes

In Chapter 5 linetypes were created which contained text items. It was also stated that complex linetypes could be customized to contain shapes. Now that we have discussed shape creation, we will investigate how complex linetypes containing shape items can be created by making four new linetypes. These new linetypes will contain one shape item from each of our four shape files.

The descriptor for complex linetypes containing shapes has the format:

*NAME, description
A,pen moves,[SHAPENAME,shape file name,S,R,X,Y],pen moves.

Note

a) The SHAPENAME must be the same as the shape name in the .SHP file.

b) The shape file name is the **compiled** (i.e .shx) name.

c) The S,R,X,Y are the same variables as used previously.

1 Open your STDA3 template file.

2 Compile and load the four written shape files, i.e. Test.shp/shx, Myshp1.shp/shx, Myshp2.shp/shx and Myshp3.shp/shx.

3 Activate Notepad and enter the following lines (remember <R>):
   *LSHPLINE,——[LSHP]——[LSHP]
   A,5,–2,[LSHP,test.shx,X=–0.25,Y=2],–10
   *CLIPLINE,——[CLIP]——[CLIP]
   A,6,–2,[CLIP,myshp1.shx,X=–0.25,Y=–1],–6
   *CONCLINE,——[CONC]——[CONC]
   A,5,–2,[CONC,myshp2.shx,R=90,X=–0.25,Y=–1],–5
   *CIRCLINE,——[CIRCS]——[CIRCS]
   A,5,–2,[CIRCS,myshp3.shp,X=6,Y=0],–7

4 *a*) File–Save As with the file name r14cust\MYLINSHP.SHP
   b) File–Exit to return to AutoCAD.

5 At the command line enter -**LINETYPE <R>** and:
   a) enter **L <R>** – the load option
   b) enter *** <R>** – all linetypes
   c) pick **r14cust\MYLINSHP** then Open
   d) pick **r14cust\test** shape file then Open
   e) escape key to end command.
   Note: the prompt as (*d*) surprised me – I did not expect it!

6 Menu bar with Format–Layer and make four new layers with a new linetype set to each:
layer	*linetype*
L1	lshpline
L2	clipline
L3	concline
L4	circline.

7 Making each new layer current, draw a line, circle and polyline shape to display the new linetypes – Fig. 8.6.

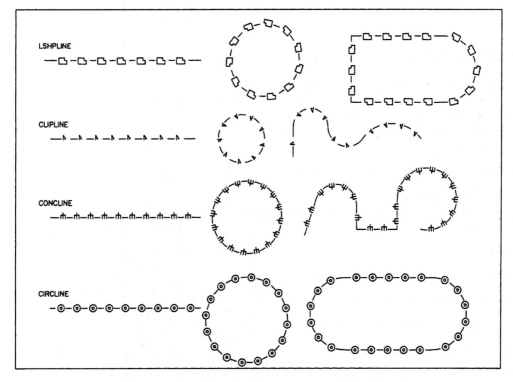

Figure 8.6 Complex linetypes containing shapes from user-written .SHP files.

8 Remember that you may have to alter the LTSCALE value and the ltScale value of individual lines for optimum effect.

9 Save if required, this exercise is now complete.

Summary

1 Shape creation and usage involves:
   a) drawing the shapes on paper with drawing unit sizes
   b) coding the shapes
   c) creating the ABC.SHP file using a text editor
   d) compiling the shape file ABC.SHP file into the ABC.SHX file
   e) loading the compiled ABC.SHX file
   f) using the SHAPE command.

2 The shape definition descriptor is:
   *Number,Bytes,NAME
   coded definitions.

3 The three commands for use with shapes are all entered at the command line, i.e.
   a) COMPILE: to compile the ABC.SHP file
   b) LOAD: to load the compiled ABC.SHX file
   c) SHAPE: to activate the shape command.

4 Special codes exist for specific use, some of which are:
   001 (or 1): pen down mode, i.e. draw
   002 (or 2): pen up mode
   008 (or 8): single *X,Y* vector displacement
   009 (or 9): multiple *X,Y* vector movements
   00A (or 10): curved vectors – octantarcs.

5 **All** shape definitions must end with 0.

6 The 009 (or 9) code must end with 0,0.

7 The number of bytes in the first line of the shape definition file must equal the number of bytes in the second line.

8 Shape numbers in any file must be between 1 and 255.

9 The pen is always assumed to be down.

10 When an existing shape file has been modified, it is necessary to exit AutoCAD then re-open the drawing before compiling the modified shape file.

11 It is recommended that all shape files are 'stored' in the same folder as the drawings in which they are used.

12 When a drawing containing shapes is opened, AutoCAD 'looks' for the corresponding shape file.

Assignment

For the shape activity we will return to the computer shop. One of the systems displays a games package and you have to add some 'pixels' to the screen as shapes.

Activity 7: Computer pixel shapes

1 Open the original computer layout r14cust\WINDOW of the six computer shapes with attribute information.

2 Erase all the attribute data and five of the computer symbols then scale the remaining symbol to suit.

3 Create the five shapes in a file r14cust\COMP.SHP. The shape names and drawing units sizes are given.

4 Compile and load the new shape file.

5 Use the SHAPE command to insert the five new shapes on the screen using your imagination for the layout.

6 Save as r14cust\COMPSHP when complete.

Chapter 9

Slides

When a drawing is 'opened' a file with the extension .dwg is displayed. This file may contain information about dimensions, layers, blocks, attributes, text styles, etc. and all this data must be processed before the drawing can be viewed on the screen. This takes time and uses memory.

AutoCAD has a facility which can 'capture' drawings as photographic snapshots called **slides** with several advantages:

1 They are easy to make.

2 They can be quickly viewed.

3 They do not use a lot of memory.

4 They allow several 'pictures' of the one drawing to be stored for future recall.
Slides are files with the extension **.SLD** and can be stored on a floppy disk or in a named folder.

Uses for slides

Slides are 'pictures' of the drawing screen and once created, **cannot be modified**. They can be used for many purposes, some of which are:

1 Running a slide show.

2 Making a slide library.

3 Creating hatch pattern icons.

4 Creating icons for use with menus.

5 Simple animation.

6 Presentation work.

7 Project work.

In this chapter we will investigate several of these slide uses with different worked examples.

Note

1 While slides are very easy to make, the process is repetitive and can become rather tedious.

2 Slides are generally used with script files (next chapter), so their full potential will not be evident in this chapter.

3 When creating slides, it is advisable to take a note of the slide names. It is very easy to forget what slides have been created.

Slide example 1 – slides for a slide show

A slide show is a series of slides 'run together' using a script file. To run a slide show, the slides must obviously have been prepared and our example will consider a component being constructed for a pressing operation, so:

1 Open your STDA3 template file with layer OUT current.

2 Refer to Fig. 9.1 which gives the stages in the construction of the slides.

Creating the slides

1 *a*) Draw the component using the sizes given in Fig. (a). Ensure that the lower left corner
       is positioned at the point (50,50).
    b) Add the given dimensions on layer DIM.
    c) Add the item of text on layer TEXT.

2 At this stage, save the drawing as **r14cust\SHOW**.

3 Freeze layer 0.

4 At the command line enter **MSLIDE <R>** and:
    prompt   *Create Slide File dialogue box*
    respond   1. ensure **r14cust** current
                2. ensure Slide (*.sld) is file type
                3. enter File name as **SL1**
                4. pick Save.

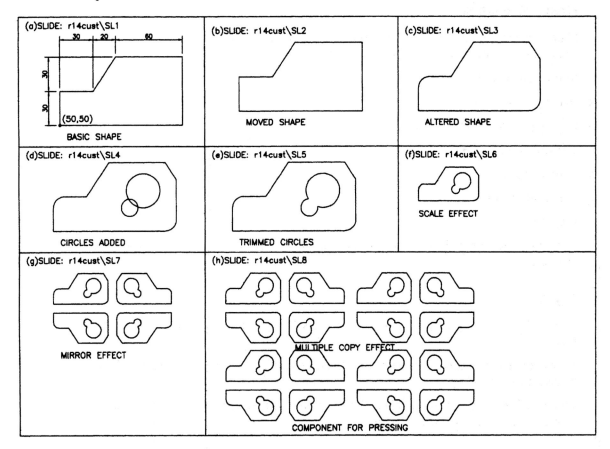

Figure 9.1 Stages in constructing slides SL1–SL8 for slide show.

5 *a*) Erase the dimensions.
   b) Move the shape and text item from 50,50 to 20,220.
   c) Alter the text item as Fig. (b).

6 At the command line enter **MSLIDE <R>** and:
   a) Create Slide File dialogue box with r14cust current?
   b) Enter slide file name as SL2.
   c) Pick Save.

7 *a*) Fillet and chamfer any 2/3 corners using your own values.
   b) Modify the text – Fig. (c).
   c) Make a slide with file name r14cust\SL3.

8 *a*) Add two intersecting circles on layer OUT and alter the text item – Fig. (d).
   b) Make a slide with file name SL4.

9 *a*) Trim the intersecting circles.
   b) Alter the text item – Fig. (e).
   c) Make a slide with name SL5.

10 *a*) Scale the shape about the point 20,280 by a factor of 0.6667.
   b) Move and alter the text – Fig. (f).
   c) Make a slide – SL6.

11 *a*) Mirror the scaled shape about a vertical line.
   b) Mirror the two shapes about a horizontal line
   c) Move and alter the text – Fig. (g).
   d) Make a slide – SL7.

12 *a*) Multiple copy the four shapes to three other places on screen.
   b) Move and alter the text – Fig. (h).
   c) Thaw layer 0.
   d) Add a title and make a slide – SL8.

Viewing the slides

The created slides can be viewed with:

1 At command line enter **VSLIDE <R>** and:
  prompt   Select Slide File dialogue box
  respond   1. ensure r14cust is current
            2. pick SL1
            3. pick Open.

2 The first slide of the basic shape with dimensions will be displayed.

3 Repeat the command line VSLIDE entry, and pick the other slides SL2–SL8 to display each created slide.

4 When the eight slides have been displayed, either:
   a) menu bar with **View–Regen** (or View–Redraw)
   b) select the REDRAW icon from the Standard toolbar
   c) enter REGEN <R> or REDRAW <R> at the command line. This will 'return' the drawing screen to the original display prior to the VSLIDE command.

Note

1 Slides are 'raster images' of the graphics screen and cannot be altered. View any of your slides SL1–SL8 and try and erase any object – you cannot.

2 There are only three commands used with slides:
   a) MSLIDE: to create the slides
   b) VSLIDE: to view the slides
   c) REGEN/REDRAW: to refresh the drawing screen.

3 The three commands are entered from the command line.

4 At present this is all that can be achieved with slides.

Slide example 2 – slides for an animation

An animation is similar to a slide show as it requires both slides and a script file to be created. The example selected for the animation is in 3D and should be quite interesting to the user.

1 *a*) Open your STDA3 template file with layer OUT current.
   b) Erase the border, grid off.
   c) Menu bar with **View–3D Viewpoint–SE Isometric**.

2 Menu bar with **Draw–Surfaces–3D Surfaces** and:
   prompt   3D Objects dialogue box
   respond **pick Dish then OK**
   prompt   Center of dish and enter: **100,100 <R>**
   prompt   Diameter/<radius> and enter: **75 <R>**
   prompt   Number of longitudinal segments and enter: **32 <R>**
   prompt   Number of latitudinal segments and enter: **16 <R>**.

3 A red dish will be displayed.

4 Menu bar with Draw–Surfaces–3D Surfaces and:
   a) pick Sphere from the list then OK
   b) center of sphere: 25,100,25
   c) radius of sphere: 25
   d) longitudinal segments: 16
   e) latitudinal segments: 8.

5 With the PROPERTIES icon from the Standard toolbar, change the colour of the sphere to blue.

6 The basic layout is as Fig. 9.2.

7 At this stage save the drawing as **r14cust\ANIM**.

8 At the command line enter:
   a) HIDE <R> to display the layout with hidden line removal
   b) MSLIDE <R> and make a slide with file name r14cust\ANIM1.

9 With the ROTATE icon from the Modify toolbar:
   a) Select object: pick the blue sphere then right-click
   b) Base point: enter 100,100 <R>
   c) Rotation angle: enter 20 <R>.

10 At command line enter:
   a) HIDE <R>
   b) MSLIDE <R> with file name r14cust\ANIM2.

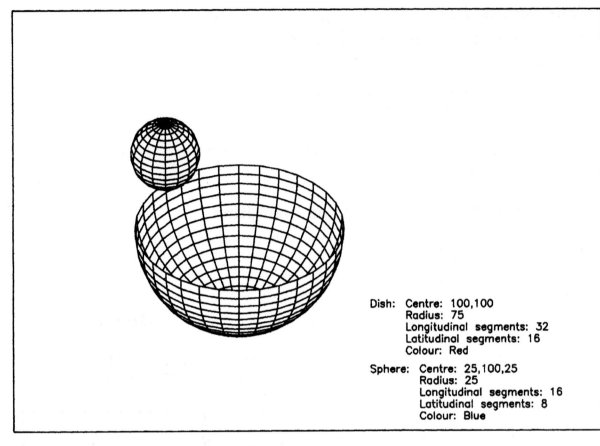

Dish: Centre: 100,100
Radius: 75
Longitudinal segments: 32
Latitudinal segments: 16
Colour: Red

Sphere: Centre: 25,100,25
Radius: 25
Longitudinal segments: 16
Latitudinal segments: 8
Colour: Blue

Figure 9.2 Original layout for animation slides.

11 Repeat steps 9 and 10 to make another 16 slides, i.e.
   a) rotate the sphere about the point 100,100 for 20 degrees
   b) hide the model
   c) make a slide r14cust\ANIM3, r14cust\ANIM4 – r14cust\ANIM18
   d) told you creating slides could be boring.

12 At present the slides can be viewed, but the animation cannot be prepared until we have investigated script files.

Slide example 3 – slides for a slide library

A slide presentation may contain dozens of slides and this can result in folders becoming congested with .sld files. In our two examples we have already created 26 slide files in the r14cust folder and will create more before the chapter is finished.

When working with slides it is useful to 'store' all related slides in a *slide library*. To create a slide library it is necessary to use a utility program called **SLIDELIB** supplied with AutoCAD. To demonstrate how a slide library is created we will prepare a new set of slides – the construction of a 2D flange drawing.

Creating the slides

1 Open your STDA3 template file and refer to Fig. 9.3.

2 With layer CL current, draw two centre lines:
   a) horizontal, from: 20,140; to: @340,0
   b) vertical, from: 120,240; to: @0,–200 – Fig. (a)
   c) make a slide with file name: r14cust\FL1.

3 With layer OUT current:
   a) draw three concentric circles at the centre line intersection with radii 90, 50 and 30
      – Fig. (b)
   b) make a slide with file name FL2 – r14cust current?

4 Add a 'circular bolt hole', centre at 120,210 with radius 10 and make a slide FL3 –
   Fig. (c).

5 Polar array the red bolt hole:
   a) about any circle centre
   b) for 4 items, 360 angle with rotation
   c) make a slide FL4 – Fig. (d).

6 With layer 0 current, draw five horizontal construction lines of length 240 and make a
   slide FL5 – Fig. (e).

7 With layer OUT current, draw the top half of the right-hand view (five lines), the 'widths'
   being 40 and 70 – Fig. (f). Make a slide named FL6. Use your discretion for the 'start
   point'.

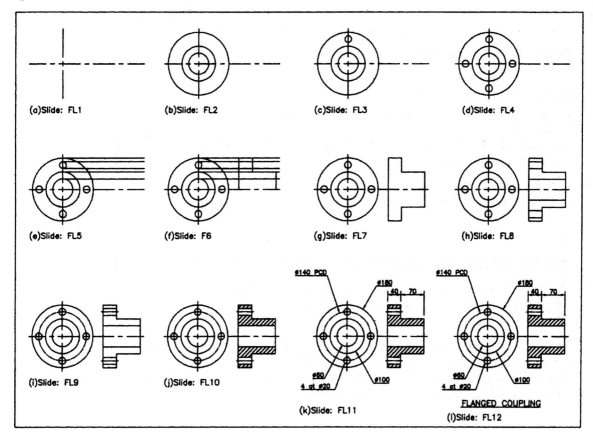

Figure 9.3 Stages in constructing slides FL1–FL12 for slide library.

8   *a*) delete the five construction lines

    b) mirror the five red outlines about the horizontal green centre line, without deleting old objects – Fig. (g)

    c) make a slide FL7.

9   Add to the right view six horizontal 'hole' lines as Fig. (h) and make a slide FL8.

10 With layer CL current:

    a) add the two bolt hole centre lines

    b) add the PCD circle – Fig. (i)

    c) optimize the LTSCALE system variable

    d) make a slide FL9.

11 With layer SECT current, hatch the four 'areas' with a user-defined pattern, angle 45, spacing 4 – Fig. (j). Make a slide FL10.

12 Add the dimensions as Fig. (k) on layer DIM and make a slide FL11.

13 Finally with layer TEXT current, add a title to the drawing as Fig. (l) and make a slide FL12.

14 At this stage save the complete drawing as **r14cust\FLANGE**.

15 View your slides if required.

The SLIDELIB utility program

AutoCAD 'stores' the slide library utility program in the SUPPORT 'sub-folder' of the AutoCAD R14 folder. I always recommend that this program is copied into the same folder as the slides and drawings, in this case r14cust. To achieve this we will activate the MS-DOS prompt command from the Windows taskbar. Some users may know of other ways to copy a program from one folder to another. I have selected the MS-DOS Prompt method for convenience.

1   Complete flange drawing still displayed?

2   Using the Windows taskbar select **Start–Programs–MS-DOS Prompt** (or Command Prompt) and:

prompt	*MS-DOS text screen*
with	C:\WINDOWS or C:\ displayed
enter	**CD\AutoCAD R14\SUPPORT <R>**
where	*a*) AutoCAD R14 is **MY** installed folder name
	b) SUPPORT is a folder within AutoCAD R14
prompt	*C:\AutoCAD R14\SUPPORT*
enter	**dir *.exe <R>**
and	list of EXE files displayed
with	slidelib.exe listed?
then	*C:\AutoCAD R14\SUPPORT* displayed
enter	**COPY SLIDELIB.EXE C:\r14cust <R>**
prompt	*1 file(s) copied*
and	command prompt returned.

3   To ensure that the slidelib utility program has been copied into our working folder:

    a) enter: CD\r14cust <R>

    b) enter: dir *.exe <R>

    c) slidelib.exe displayed.

4   Return to AutoCAD by entering **EXIT <R>**.

Creating the slide library for the flange slides

1 From AutoCAD, activate Notepad using the Windows taskbar.

2 Enter the following lines of text, remembering <R> at the end of each line:
FL1.SLD
FL2.SLD
FL3.SLD, etc. until **FL12.SLD**.

3 Menu bar with **File–Save As** and:
a) ensure r14cust folder name current
b) enter file name: **FLANGE.TXT**
c) pick Save.

4 Exit Notepad to return to AutoCAD.

5 Activate the MS-DOS PROMPT from the Windows taskbar and:
prompt   *C:\WINDOWS or C:\displayed*
enter   **CD\r14cust <R>**
prompt   *C:\r14cust*
enter   **SLIDELIB FLANGELIB<FLANGE.TXT <R>**
prompt   *SLIDELIB 1.2 (date)*
             © Copyright ... AutoDESK Inc.

6 At the command prompt:
enter   **dir *.slb <R>**
prompt   *flangelib.slb with 43916 bytes (or similar).*

7 At the command prompt enter **EXIT <R>** to return to AutoCAD.

8 *Notes.*
At this stage some explanation of what has been achieved may help the user.
a) we have created a FLANGE.TXT file of the 12 slides
b) this text file has been 'converted' into a slide library file with the extension .slb using the SLIDELIB utility program
c) the entered line SLIDELIB FLANGELIB<FLANGE.TXT can be read as:

SLIDELIB: the name of the utility program being 'run'
FLANGELIB: the name of the slide library being created
FLANGE.TXT: the source text file containing the slides
<:   a directional indicator meaning 'transfer all data from the file FLANGE.TXT into the file named FLANGELIB
.SLB: the extension .SLB is automatically added during the operation and is a compiled extension name, i.e. FLANGELIB.SLB is a compiled version of FLANGE.TXT

d) the file name FLANGELIB has been used to be 'meaningful', i.e. FLANGE: to indicate the flange slides and LIB: to indicate a slide library.

Viewing slides from a slide library file

1 Flange drawing on screen?

2 At the command line enter **FILEDIA <R>** and:
prompt New value for FILEDIA<1>
enter **0 <R>**.

3 At the command line enter **VSLIDE <R>** and:
prompt Slide file<.>
enter **FLANGELIB(FL8) <R>**.

4 The slide of the two views with hatching should be displayed.

5 At the command line:
a) enter: VSLIDE <R>
b) slide file and enter: FLANGELIB(FL5)
c) slide of left view with construction lines displayed.

6 *a*) REGEN the screen to restore the original drawing
b) reset the FILEDIA variable to 1.

7 *Note.*
a) The variable FILEDIA controls the dialogue box display. If set to 1 (the default) dialogue boxes will be displayed. If set to 0, then command line entry is required. I found it necessary to set FILEDIA to 0 to display the slides from the slide library
b) the entry FLANGELIB(FL8) can be 'read' as: displayed slide FL8 from the slide library FLANGELIB
c) a slide library requires:
　　i)　a series of slides
　　ii)　a text file listing the slides
　　iii)　the SLIDELIB utility program to convert the .TXT slide file into a .slb slide library file
d) although the slides are 'in' the slide library file, **never delete** them from your folder. They must be 'as made'
e) it is recommended that the slides, .TXT slide file, the utility program and the .slb slide library file are all in the same folder – in our case r14cust.

8 This completes the slide library exercise, although the procedure will be used with the next two slide examples.

Slide example 4 – slides for use in an icon menu

Slides can be created for use in a icon (palette) menu. Menus will be investigated in another chapter, but we will prepare the slides in this exercise, so:

1 Open your STDA3 template file and refer to Fig. 9.4 which displays information for five house icons to be used in a housing estate layout.

2 With layer OUT current, use the given sizes and draw the four icons using your discretion for any omitted dimensions. Position the icons anywhere on the screen.

3 Make blocks of the five icons:
   a) using the given names, e.g. SEMI, 3BED, etc.
   b) with the insertion base point at the top left corner of each icon.

4 With the BLOCK or INSERT command, use the ? option and check the five blocks have been defined.

5 At this stage save the drawing as **r14cust\ESTATE**.

6 Insert the block SEMI:
   a) at any suitable point on the screen
   b) full size with 0 rotation.

7 Zoom-in on the inserted block.

8 Make a slide of the screen (the zoomed area) with the file name **r14cust\ESTSL1**.

9 Erase the inserted SEMI block and zoom previous.

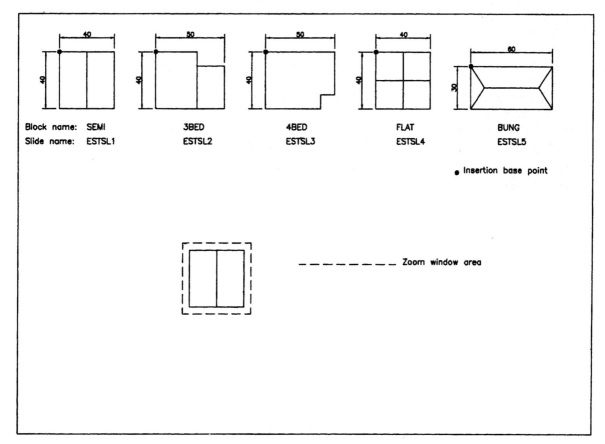

Figure 9.4 Block formation information for icon slides.

10 Repeat steps 6–8 and:
   a) insert each block at a suitable point
   b) zoom in on the inserted block
   c) make a slide of the zoomed area with the following slide names:

block	slide name
3BED	ESTSL2
4BED	ESTSL3
FLAT	ESTSL4
BUNG	ESTSL5

   d) erase the inserted block, zoom previous then repeat.

11 Activate Notepad from the Windows taskbar and:
   a) enter the following (remember <R>):
   ESTSL1.SLD
   ESTSL2.SLD
   ESTSL3.SLD
   ESTSL4.SLD
   ESTSL5.SLD
   b) save the entered file as **r14cust\ESTATE.TXT**.

12 Activate the MS-DOS Prompt from the Windows taskbar and:
   a) change to r14cust with **CD\r14cust**
   b) execute the slide library program with the following entry:
   SLIDELIB ESTATELIB<ESTATE.TXT <R>.

13 At present we cannot continue with the slides created in the slide library until we have investigate how menus are customized.

Slide example 5 – slides for use with hatch pattern icons

This example is very similar to slide example 4. We will create a series of slides which will allow our created hatch patterns from Chapter 7 to be selected from icons. The icon slides cannot be used until menus have been discussed, but the slides and slide library will have been created.

1 Open the STDA3 **drawing** file and refer to Fig. 9.5.

2 Draw a rectangle 100 × 70 and copy it to four other areas.

3 At the command line enter **HATCH <R>** and:
   prompt   *Pattern* and enter: HANG
   prompt   *Scale* and enter: 1
   prompt   *Angle* and enter: 0
   prompt   *Select objects* and pick the first rectangle
   Note: if the hatch pattern is invalid, check that you have opened your STDA3 drawing file and not the template file.

4 Hatch the other four rectangles with the hatch patterns DBOX, WEAVE, ELLS and ARROWS using the scale factor and angle values given in Fig. 9.5.

5 Zoom in on the HANG rectangle and make a slide of the screen:
   a) with r14cust current
   b) file name: **HATSL1**.

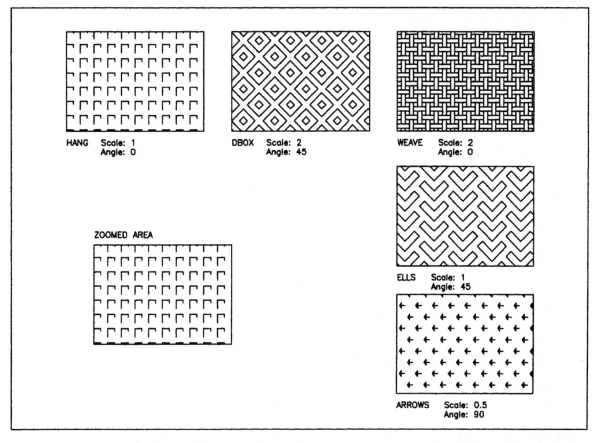

HANG   Scale: 1
        Angle: 0

DBOX   Scale: 2
        Angle: 45

WEAVE   Scale: 2
        Angle: 0

ZOOMED AREA

ELLS   Scale: 1
        Angle: 45

ARROWS   Scale: 0.5
         Angle: 90

Figure 9.5 Information for creating hatch icon slides.

6 Zoom–previous and repeat step 5 for the other hatched rectangles, the slide names being:

hatch pattern	*slide name*
DBOX	HATSL2
WEAVE	HATSL3
ELLS	HATSL4
ARROWS	HATSL5.

7 Activate Notepad from the Windows taskbar and:
   a) enter the following (remembering <R>):

   HATSL1.SLD
   HATSL2.SLD
   HATSL3.SLD
   HATSL4.SLD
   HATSL5.SLD

   b) save the entered file as **r14cust\HATCH.TXT**.

8 Activate the MS-DOS Prompt from the Windows taskbar and:
   a) change to r14cust with **CD\r14cust <R>** – if required
   b) execute the slide library program with the following entry:
   SLIDELIB HATCHLIB<HATCH.TXT <R>

9 At present we cannot continue with the slides created in the slide library until we have investigate how menus are customized.

10 At the r14cust prompt, enter **dir *.slb** to display the slide libraries which have been created. There should be three:
FLANGELIB
ESTATELIB
HATCHLIB.

11 Exit back to AutoCAD.

12 This completes the slide chapter.

Summary

1 Slides are 'snapshots' of the drawing screen.

2 Slides are raster images and cannot be modified when created, but they can be 'redefined'.

3 The slide commands are entered from the keyboard and are:
MSLIDE: to create a slide, the user entering the slide file name
VSLIDE: to view a saved slide
REGEN\REDRAW: to restore the original drawing screen.

4 Slides should be 'stored' in the same folder as the drawing from which they are created.

5 Related slides can be stored in a slide library.

6 A slide library requires:
a) a series of created slides
b) a user-written text program of the slides
c) the utility program SLIDELIB.

7 For ease of use, it is recommended that the SLIDELIB utility program is copied into the active folder.

8 Slides have many uses, but their full potential will not be realized until we have investigated script files and menus.

Assignment

I have not included any activity for this chapter, as the exercises have been rather involved. Hopefully the user will have grasped how to create and view slides and (more importantly) how to create slide libraries using SLIDELIB.

Chapter 10

Script files

Script files are text files and are used:
a) to run slide shows and animations
b) display drawing routines
c) execute frequently used routines.

Script file commands

Script files are written by the user using a text editor (Notepad) and have the extension
.SCR. With slide shows and animations, the commands used in a script file are:

VSLIDE xxx   display the slide named xxx
DELAY *nnn*   gives a pause between lines in a scrip file. The delay is *nnn milliseconds*
and the user specifies the value of *nnn*
RSCRIPT   returns to the first line of the script file, i.e. it reruns (cycles) the
complete file. The ESC key will stop a script file.

Script example 1 – a slide show

1 Open the saved drawing **r14cust\SHOW** from Chapter 9 to display the shape used to
create the slides.

2 Activate Notepad from the Windows taskbar and:
a) enter the following lines (remember <R>)
   VSLIDE SL1
   DELAY 2000
   VSLIDE SL2
   DELAY 2000
   VSLIDE SL3
   DELAY 2000
   VSLIDE SL4
   DELAY 2000
   VSLIDE SL5
   DELAY 2000
   VSLIDE SL6
   DELAY 2000
   VSLIDE SL7
   DELAY 2000
   VSLIDE SL8
   DELAY 2000
   RSCRIPT
b) Menu bar and:
   i) ensure r14cust is current folder
   ii) enter file name: **SHOW.SCR**
   iii) pick Save
   iv) return to AutoCAD (exit Notepad or minimize screen).

3   Menu bar with **Tools–Run Script** and:
    prompt   Select Script File dialogue box
    respond   1. ensure r14cust current and *.scr is file type
             2. pick **show** then Open.

4   The eight slides will be displayed with a short pause between each one, allowing the user
    to 'describe' to the audience what is happening.

5   When you have had enough of the slide show, press the **ESC** key.

6   Regen or redraw to restore the original drawing screen.

7   *Note.*
    a) The delay number can be increased or decreased as required.
    b) DELAY 1000 gives about a 1 second pause.
    c) I found that I had to activate the Run Script selection twice as I had a VVSLIDE error.
        I could not reason out why?

8   *Task.*
    Before leaving this exercise:
    a) select from the menu bar File–New and:
        i) pick No to save changes
        ii) pick Use a Wizard–Quick Setup–Done
    b) menu bar with Tools–Run Script and:
        prompt   Select Script File dialogue box
        respond   1. change to r14cust folder
                 2. pick file **show** then Open
        prompt   AutoCAD Text Window
        with   'Can't find file in search path' message with a list of named folders
        respond   cancel the command and the text window
    c) to run a script file, it is necessary to open a drawing from the folder containing the
        slides, i.e. r14cust.

Script example 2 – an animation

This example is basically the same as the first, but the script file has no delays in it,
thereby giving the 'continuous' motion effect of an animation.

1   Open the saved drawing **r14cust\ANIM** from Chapter 9.

2   Activate Notepad and:
    a) enter the following lines (remember <R>):
        VSLIDE ANIM1
        VSLIDE ANIM2
        VSLIDE ANIM3
        VSLIDE ANIM4, etc. until
        VSLIDE ANIM17
        VSLIDE ANIM18
        RSCIPT
    b) save the entered file as **r14cust\ANIM.SCR** and return to AutoCAD.

3   Erase the drawing from the screen.

4 At the command line enter **SCRIPT <R>** and:
   prompt   *Select Script File dialogue box*
   respond   1. ensure r14cust current
              2. pick anim then Open.

5 The screen will display the animation of the blue ball rotating about the red dish – ESC to stop animation.

6 *Note.*
   a) Again I had to activate the SCRIPT command twice due to a VVSLIDE error.
   b) The animation effect was not what I expected – it was poor.
   c) I have had better effects using R12 and R13 for this animation, perhaps your result was better than mine?
   d) A slide show and an animation are the same, the only difference being in the value of the delay between the slides.

A variation on slide shows

The two script files which have been written have used the commands VSLIDE, DELAY and RSCRIPT. While these are basically all that is required to run a slide show/animation, there is a variation which we will discuss using the script example 1 file as our example. This file could have been written as:

VSLIDE SL1	Begin show and load slide SL1
VSLIDE *SL2	Preload slide SL2
DELAY 2000	View slide SL1
VSLIDE	Display slide SL2
VSLIDE *SL3	Preload slide SL3
DELAY 2000	View slide SL2
VSLIDE	Display slide SL3
VSLIDE *SL4	Preload slide SL4
DELAY 2000	View slide SL3, etc.

It is the user's preference as to whether to preload slides before the DELAY command. The preload routine is considered 'better practice', but I generally do not use it.

Script example 3 – a drawing routine

Script files can be written which will produce a drawing. To demonstrate the concept:

1   Open your STDA3 **drawing** file.

2   Activate Notepad and:
   a) enter the following lines and:
     i)   ensure that there is NO SPACE between the last entry in a line and the <R>
     ii) the ^ symbol represents a press on the spacebar
   ;a drawing sequence<R>
   COLOUR^RED<R>
   PLINE^10,10^W^10^10^410,0^410,290^10,290^C<R>
   COLOUR^BLUE<R>
   DONUT^120^140^120,200^280,200^200,130<R>
   <R>
   DELAY^5000<R>
   CHANGE^L^^P^C^3<R>
   ERASE^C^5,5^400,290<R>
   REDRAW<R>
   b) File–Save As **r14cust\DRAW.SCR**
   c) Exit back to AutoCAD (or minimize the screen).

3   Activate the script command and:
   a) select the file: r14cust\Draw
   b) pick Open.

4   The drawing screen will display:
   a) a red polyline border
   b) three blue donuts
   c) the last donut will change colour to green
   d) the polyline and donuts will be erased
   e) the original drawing screen will be returned.

5   *Notes.*
   a) This script file uses the AutoCAD commands as they would be used from the keyboard.
   b) Coordinates are specified as 10,10.
   c) Spaces in lines (donated by ^) are equivalent to <R> in a command sequence.
   d) A blank line (<R>) signifies the use of the <RETURN> key at the end of the previous line – think about this!
   e) The donut line is thus:

DONUT	the AutoCAD command
space	<R> after entering DONUT
120	the donut inner diameter
space	<R> after entering 120
140	the donut outer diameter
space	<R> after entering 140
120,200	coordinates of donut centre
space	<R> after entering 120,200
280,200	coordinated of next donut centre
space	<R> after entering 280,200
200,130	coordinates of next donut centre
space	<R> after entering 280,200
<R>	end of line return
<R>	the next line <RETURN> is equivalent to cancelling the donut command.

f) To write lines in this type of file, the user records every entry for each command to be used, i.e. you try the sequence manually from the keyboard, noting each step as it happens.

g) Comments can be added using the semi-colon (;) at the start of a line – as the first line in the file.

Script example 4 – a user routine

Frequently used routines can be made into script files and run at any time. To demonstrate this, we will write a file to 'set' an A4 standard sheet with layers, so:

1 Begin a New drawing and:
  a) Start from Scratch
  b) Metric then OK.

2 Activate Notepad and enter the following lines remembering:
  a) no spaces between the last text item entered and <R>
  b) the ^ symbol indicates a press on the spacebar

```
;A4 SETUP<R>
ERASE^ALL<R>
<R>
BLIPMODE^OFF<R>
UNITS^2^2^1^0^0^N<R>
GRID^10<R>
SNAP^5<R>
LAYER^M^OUT^N^CL,HID,DIM,TEXT<R>
<R>
LAYER^C^1^OUT^C^3^CL^C^4^HID^C^5^DIM^C^6^TEXT<R>
<R>
LAYER^L^CENTER^CL^L^HIDDEN^HID<R>
<R>
LIMITS^0,0^297,200<R>
LINE^0,0^285,0^285,190^0,190^C<R>
ZOOM^A<R>
```

3 Save this file as: **r14cust\A4SETUP.SCR**.

4 Exit back to AutoCAD.

5 Activate the script command and select the file **r14cust\a4setup**.

6 The script file will:
  a) erase all objects from the screen
  b) turn blips off
  c) set the grid and snap to 10 and 5 respectively
  d) create five new layers with layer OUT current
  e) set the layer colours and linetypes
  f) set the drawing limits to A4 size
  g) draw a border then zoom all.

7 *Task*
  a) check the layer dialogue box
  b) the A4SETUP script file draws the border in red on layer OUT.
  Can you modify the script file so that the line border is on layer 0?

8 This completes the script exercises.

Summary

1 Script files are text files written by the user.

2 Script files are used for:
   a) slide shows and animations
   b) programmed drawing routines
   c) frequently used 'operations'.

3 Script files are activated:
   a) from the menu bar with Tools–Run Script
   b) with SCRIPT at the command line.

4 It is recommended that script files are stored in the working folder.

Chapter 11

Customizing menus

AutoCAD Release 14 has a very comprehensive user-friendly menu structure, and the standard menu layout should be sufficient for all users most of the time. An occasion may arise when it would be advantageous to alter the menu structure to suit your own 'advanced' skills or to meet a customers requirements.

There are different 'types' of menus within AutoCAD and users will be familiar with most of them, without (perhaps) knowing their names. The menu types available include:
a) the screen menu – not always used
b) pull-down menus – from the menu bar
c) pop-up menu dialogue boxes
d) cascade menus
e) icon (image tile) menus – hatch patterns
f) tablet menus
g) toolbars
h) keyboard accelerators
i) pointing device button menus
j) auxiliary menus.

Menus are text files written by the user using a text editor (Notepad) and menu customization can be achieved in two ways:

1   by altering the existing AutoCAD menu

2   by creating a new menu.

In this chapter we will investigate several of the menu types listed above by creating our own customized menu, building it up in stages as more options are considered. This will allow the existing menu system to remain untouched – a wise precaution?

Note: this chapter is rather long. It is advisable to 'work through' all of the menu exercises so that no part of the menu structure is missed out.

AutoCAD R14's menus

To investigate the existing menu structure within AutoCAD:

1 Start up and open your STDA3 template file.

Note: in the following exercise the user should know the folder name into which AutoCAD has been installed. I have assumed it is AutoCAD R14.

2 Activate the MS-DOS Prompt from the Windows taskbar and:
  prompt C:\ or similar
  enter   **CD\AutoCAD R14\SUPPORT <R>**
  prompt *C:\AutoCAD R14\SUPPORT*
  enter   **dir acad.mn? <R>**
  and`   list of AutoCAD menus with dates, times and file sizes
            acad.mnc
            acad.mnr
            acad.mns
            acad.mnl
            acad.mnu.

3 *Notes*:
  a) AutoCAD has several menu types, the file extensions being:
     .mnu: template menu file written by the user
     .mnc: compiled menu file used by AutoCAD. It contains the information necessary
           for the user to use the menu system
     .mnr: resource menu file containing the bitmaps used by the menu
     .mns: source menu file modified by AutoCAD
     .mnl: menu LISP file containing AutoLISP expressions.
  b) The only menu type that the user is concerned with is the **.mnu** template file. This
       file is written by the user. AutoCAD compiles the .mnc and .mnr files itself, and
       modifies the .mns file as a drawing is being created.

4 Cancel the MS-DOS Prompt screen and return to AutoCAD.

First menu – a simple screen menu

Most users will probably be using Release 14 without a screen menu, but the facility exists within R14 to display the screen menu. The first menu example will create a simple screen menu and we will write it, use it and then discuss its contents.

1 In AutoCAD with the STDA3 template file displayed?

2 Cancel all toolbars.

3 Menu bar with **Tools–Preferences** and:
  prompt   *Preferences dialogue box*
  respond 1. pick Display tab
             2. pick Display AutoCAD screen menu in drawing window, i.e. tick in box
             3. pick OK.

4 The screen menu will be displayed at the right-hand side of the monitor screen with the same options as the menu bar.

5 Activate Notepad from the Windows taskbar and:
   a) enter the following lines of text but note:
       i)   spaces are entered with a spacebar press
       ii)   the line numbers are for reference only
       iii) remember <R> **immediately** after each entry

***SCREEN	line 1
[MYMENU]	line 2
<R>	line 3
[STRAITS]^C^C_LINE	line 4
[ROUNDS]^C^C_CIRCLE	line 5
[WORDS]^C^C_DTEXT	
<R>	
[RUBOUT]^C^C_ERASE	
<R>	
[ZOOMIN]^C^C_ZOOM W	line 6
[ZOOMOUT]^C^C_ZOOM P	

   b) menu bar with File–Save As and:
       i)   ensure r14cust current
       ii) file name: **MYMENU.MNU**
       iii) pick Save
       iv) minimize Notepad to return to AutoCAD
   c) *note*: by minimizing Notepad we will not need to open the MYMENU.MNU file every
   time we want to alter it. We need only select Mymenu–Notepad from the Windows
   taskbar to 'enter' the MYMENU.MNU file.

6 At the command line enter **MENU <R>** and:
   prompt   *Select Menu File dialogue box*
   respond   1. ensure r14cust current, i.e. Look in
               2. alter file type to **Menu Template (*.mnu)**
               3. pick Mymenu
               4. pick Open
   prompt   *AutoCAD Message dialogue box* as Fig. 11.1 – don't panic
   respond   pick Yes.

7 The screen 'appearance' will change and:
   a) new menu displayed at right – ours
   b) the Standard and Object Property toolbars are not available
   c) the menu bar will only display File and Help
   d) a zoom all may be needed?

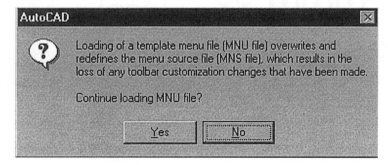

Figure 11.1 Menu file message dialogue box.

8 Use the screen menu to draw some lines, circles and text similar to Fig. 11.2(a). You have the facility to erase objects and zoom in and out of areas of the drawing. Try all screen menu options then save your drawing as **r14cust\MYDRAWG**.

9 *Note.*

   a) At present the right-hand mouse button will not work, as it has not yet been 'programmed'. You will need to use the <R> key to end commands. This is a nuisance but will be rectified with the next menu.

   b) All AutoCAD commands are still available with command line entry. Try some, e.g. MOVE, COPY, PLINE, etc.

   c) The objects drawn should be in red if the STDA3 template file was 'opened'. Why is this?

10 Exit AutoCAD with your menu still displayed.

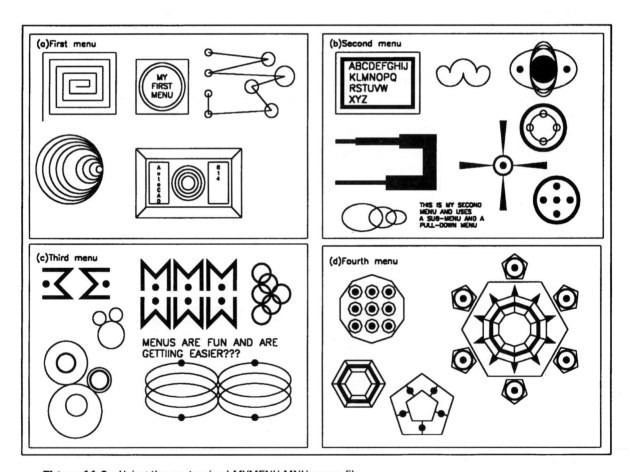

Figure 11.2   Using the customized MYMENU.MNU menu file.

Menu explanation

Line	Item	Discussion
1	***SCREEN	Signifies a screen menu to follow. The ******* are **section labels** used to 'identify' main parts of the menu
2	[MYMENU]	The menu title placed within [ ] brackets
3	<R>	Gives a blank line in the screen menu area
4	[STRAITS]^C^C_LINE	[STRAITS]: display STRAITS on screen ^C^C: cancel any active command _: a spacer only and does nothing LINE: the AutoCAD command
5	[ROUNDS]^C^C_CIRCLE	This line can be read as: Display the word ROUNDS on the screen, and when selected: *a*) cancel any active command *b*) activate the AutoCAD CIRCLE command
6	[ZOOMIN]^C^C_ZOOM W	Display the word ZOOMIN on the screen and when selected: *a*) cancel any active command *b*) activate the Zoom Window option

Notes

1 All menus are basically written as above.

2 Items placed within [ ] brackets are displayed in the screen menu area. A maximum of eight characters can be placed within [ ].

3 The ^C^C (^ is shift 6) is used to cancel any active command and has been inherited from pre-R13 versions when CTRL-C was the cancel command before the ESC key was used. The ^C^C is used **after every [ ] item**.

4 A space in a menu line is equivalent to a <R> press.

Second menu – the right button, a sub-menu and a pull down

This menu will build on the first menu by 'programming' the right button of the mouse and introducing a sub-menu and a pull-down menu.

1   Start AutoCAD R14 and open the saved **r14cust\MYDRAWG** drawing created from the first customized menu. The screen should display your customized menu – why is this?

2   Activate Mymenu–Notepad from the Windows taskbar and add the following lines to the file, noting that I have listed the complete file and added (N) for the new lines, and remember <R> after each line entry

***AUX1	N1
;	N2
***SCREEN	
**MAIN	N3
[MYMENU]	
<R>	
[STRAITS]^C^C_LINE	
[ROUNDS]^C^C_CIRCLE	
[WORDS]^C^C_DTEXT	
<R>	
[RUBOUT]^C^C_ERASE	
<R>	
[ZOOMIN]^C^C_ZOOM W	
[ZOOMOUT]^C^C_ZOOM P	
<R>	N4
[THICKS]^C^C_$S=THICKLINES	N5
**THICKLINES	N6
[POLYLINE]	N7
<R>	N8
[PLINE1]^C^C_PLINE \W 5 5	N9
[PLINE2]^C^C_PLINE \W 10 0	N10
[DONUT1]^C^C_DONUT 0 10	N11
[DONUT2]^C^C_DONUT 25 30	N12
<R>	N13
<R>	N14
<R>	N15
<R>	N16
[MAINMENU]^C^C_$S=MAIN	N17
***POP1	N18
[DRAWING]	N19
[ARC]^C^C_ARC C	N20
[DONUTUSER]^C^C_DONUT	N21
[ELLIPSE]^C^C_ELLIPSE C	N22
[PLINEUSER]^C^C_PLINE \W	N23.

3   *a*) Menu bar with File–Save to automatically update MYMENU.MNU.
    b) Minimize Notepad to return to AutoCAD.

4   Compile the new menu file with:
    a) enter MENU <R> at the command line
    b) ensure r14cust is current
    c) pick Menu Template (*.mnu) file type
    d) pick Mymenu then Open
    e) Message dialogue box – pick Yes.

5 The drawing screen will display:
   a) screen menu with STRAITS-THICKS at right
   b) selecting THICKS will display the POLYLINE sub-menu
   c) selecting MAINMENU will restore STRAITS-THICKS
   d) menu bar displays DRAWING with four selections.

6 Erase all existing objects then use the new menu items to create a layout of your choice
   – see Fig. 11.2(b).
   Note:
   a) the mouse right button now 'works'
   b) save your drawing if required.

Second menu explanation

There are several new concepts which have been added to this menu and a detailed explanation will now be given. This explanation is 'quite long', but I would advise you to read it anyway.

Programming the mouse right button

The first two lines of the second menu (N1 and N2) are new. *****AUX1** is a main section label indicating an auxiliary menu item to follow and is used for the mouse. The mouse left button is **always active** and a semi-colon (;) in the second line activates the right button – simple?. If a three button mouse is being used, you would write:

***AUX1
;   – second button active
;   – third button active

Release 12 users who have customized menus will realize that ***AUX1 has replaced ***BUTTONS for the main section label.

The sub-menu THICKLINES

Lines N4–N12 and line N3 are the sub-menu lines added to the menu.

N4: a blank line to 'space out' the word THICKS

N5: **[THICKS]^C^C_$S=THICKLINES**
[THICKS]: the word THICKS is displayed in the screen menu area
^C^C_: cancel any active command
$S=THICKLINES: 'go to' a screen sub-menu called THICKLINES.
The **$** is the sub-menu 'command' and **S** is for a screen sub-menu.
Thus when THICKS is selected from the screen menu, any active command is cancelled, and the screen sub-menu called THICKLINES will be displayed in the screen menu area.

N6–N12: the sub-menu THICKLINES

N6: ****THICKLINES**
**: the start of any sub-menu
THICKLINES: the sub-menu name
Thus $S=THICKLINES (N5)calls up the screen sub-menu THICKLINES which is activated with **THICKLINES

N7: **[POLYLINE]**
The title of the THICKLINES sub-menu

N8: A blank line to 'space out' the screen menu area

N9: **[PLINE1]^C^C_PLINE \W 5 5**
Will draw a polyline of width 5. The \ is a 'pause' which allows the user to select the polyline start point. The **W** entry activates the width option, and the entries 5 and 5 'set' the polyline width to a constant 5.

N10: **[PLINE2]^C^C_PLINE \W 10 0**
Activates the polyline command and sets the start width to 10 and the end width to 0

N11: **[DONUT1]^C^C_DONUT 0 10**
The donut command, setting diameters to 0 and 10

N12: **[DONUT2]^C^C_DONUT 25 30**
Another donut command – should be obvious?

N13–N16: four blank lines to 'blank out' the items from the main menu which would be displayed in these lines

N17: **[MAINMENU]^C^C_$S=MAIN**
The end of the THICKLINES sub-menu. When MAINMENU is selected from the screen, any active command is cancelled and then 'goto' a sub-menu called MAIN, activated with ****MAIN** – line N3. **MAIN has been 'placed' after ***SCREEN in the main menu for obvious reasons?

*The pull down menu ***POP1*

Lines N18–N23 are the first pull down menu lines.

N18:   ***POP1
          Indicates that a pull down menu is to be displayed in area 1 of the menu
          bar. *****POPn** is the section label for a pull down menu and n is the position
          in the menu bar. The lines following the ***POP1 are the items which will
          be displayed in the pull down area.

N19:   **[DRAWING]**
          The pull down title, displayed in the menu bar.

N20:   **[ARC]^C^C_ARC C**
          The word ARC will be displayed when DRAWING is selected from the menu
          bar. When arc is selected, any active command will be cancelled and the
          AutoCAD command ARC activated, with the arc centre option.

N21:   **[DONUTUSER]^C^C_DONUT**
          Activate the DONUT command, the user entering both diameters.

N22:   **[ELLIPSE]^C^C_ELLIPSE C**
          Activate the ellipse command using the Center option.

N23:   **[POLYUSER]^C^C_PLINE \W**
          Activate the polyline command, the user entering the widths.

Note

My idea with this menu was to introduce both a sub-menu and a pull-down menu at
the same time. Hopefully it has not been 'too much' for you?

The sub-menu THICKLINES allows both polylines and donuts to be drawn with 'set'
values, while the pull down menu allows the user to enter their own polyline and donut
values.

Third menu – a cascade effect and additional pull downs

In this menu we will add a cascade effect to the DRAWING pull down menu and add two additional pull down menus to the menu bar.

1 Still in AutoCAD with the second menu drawing displayed?

2 Erase all objects from the screen.

3 Activate Notepad from the Windows taskbar and:
  a) After [PLINEUSER]^C^C_PLINE \W enter:
```
[→CIRCLES]
[CENRAD]^C^C_CIRCLE
[CENDIA]^C^C_CIRCLE D
[THREE]^C^C_CIRCLE 3P
[←TTRAD]^C^C_CIRCLE TTR
***POP2
[MODIFYING]
[ERASEWIN]^C^C_ERASE W
[ERASELAST]^C^C_ERASE L
[COPY]^C^C_COPY
[MIRROR]^C^C_MIRROR
[ROTATE]^C^C_ROTATE
[MOVE]^C^C_MOVE
***POP3
[FILING]
[SAVING]^C^C_SAVE
[OPENING]^C^C_OPEN
[NEWING]^C^C_NEW
[ENDING]^C^C_QUIT
```
  b) Menu bar with File–Save to update r14cust\MYMENU.MNU
  c) Minimize Notepad.

4 Compile the altered menu with:
  a) **MENU <R>** at the command line
  b) r14cust current with Menu Template (*.mnu) file type
  c) pick Mymenu then Open and Yes to message.

5 The menu bar should display the three pull down headings with:

DRAWING	*MODIFYING*	*FILING*
ARC	ERASEWIN	SAVING
DONUTUSER	ERASELAST	OPENING
ELLIPSE	COPY	NEWING
PLINEUSER	MIRROR	ENDING
CIRCLES > CENRAD	ROTATE	
CENDIA	MOVE	
THREE		
TTRAD.		

6 Use your new menu to create a drawing, using (if possible) all the new additions – Fig. 11.2(c).

7 When complete save if required.

Third menu explanation

1 The two new pull down menus should present no problems. ***POP2 (MODIFYING) and ***POP3 (FILING) will be displayed in the menu bar next to ***POP1 (DRAWING). The selections included in these new pull downs will increase your drawing ability.

2 The new item in this menu is the cascade effect added to the ****POP1 (DRAWING) pull down. A cascade effect is obtained using the → and ← keys within the [] brackets:
[→CIRCLE]: the word CIRCLES is to be displayed in the pull down area and the (→) indicates that a follow-on will result when CIRCLES is selected. A > is displayed after the word CIRCLES. This is the cascade effect.
[←TTRAD]: displays the word THREE in cascade area and (←) ends the cascade effect.

3 When the word CIRCLES is selected from the DRAWING pull down menu, the (→) activates the cascade effect and displays CENRAD, CENDIA, THREE and TTRAD for selection. Any one of these four selections will activate the various circle options . The last entry in a cascade always has (←) to end the effect, i.e. there is a start (→) and an end (←) in a cascade menu.

Fourth menu – another two pull downs and two additional cascade menus

In this menu we will add two pull downs for dialogue boxes and object snaps, and add cascade effects for polygons and arrays.

1 In AutoCAD with drawing from menu three displayed?

2 Activate Notepad from the Windows taskbar and:
a) After [←TTRAD]^C^C_CIRCLE TTR enter:
[→MULTIS]
[PENTAGON]^C^C_POLYGON 5 \
[HEXAGON]^C^C_POLYGON 6 \
[OCTAGON]^C^C_POLYGON 8 \
[←DECAGON]^C^C_POLYGON 10 \
b) After [MOVE]^C^C_MOVE enter:
[→ARRAYS]
[RECTANGULAR]^C^C_ARRAY \ R
[←POLAR]^C^C_ARRAY \ P
c) After [ENDING]^C^C_QUIT enter:
***POP4
[BOXES]
[layers]^C^C_DDLMODES
[aids]^C^C_DDRMODES
[units]^C^C_DDUNITS
[change]^C^C_CHPROP
***POP5
[SNAPS]
[endpoint]END
[midpoint]MID
[centre]CEN
[intersection]INT
[quadrant]QUA
[perpendicular]PER
d) Menu bar with File–Save to update r14cust\MYMENU.MNU
e) Minimize Notepad to return to AutoCAD.

3   Erase all objects from and compile the modified menu with:
    a) command line **MENU <R>**
    b) ensure r14cust current with Menu Template (*.mnu) file type
    c) pick Mymenu then Open and pick Yes to message.

4   Screen displays the new menu with five items in the menu bar, BOXES and SNAPS having been added.

5   Refer to Fig. 11.2(d) and use your new menu to create another drawing – use your imagination! You have two new options:
    a) cascade effect with the DRAWING pull down of MULTIS to activate four polygon options
    b) cascade effect with the MODIFYING pull down of ARRAYS to activate the rectangular and polar array commands.

6   When complete save the drawing if required.

Fourth menu explanation

The additions to the fourth menu are really 'repeats' of what we have already entered in previous menus.

1   The pull down menu ***POP4 (title BOXES) allows the user access to four dialogue boxes – layers, drawing aids, units and change properties. The dialogue boxes have been 'called' with the dynamic dialogue commands DDLMODES (layers), DDRMODES (aids), DDUNITS (units) and DDCHPROP (properties).

2   Thus the line [layers] $^\wedge$ C $^\wedge$ C_DDLMODES can be read as:
    When [layers] is selected from the BOXES menu bar item, cancel any active command and activate the layer control dialogue box.

3   The pull down menu ***POP5 (title SNAPS) will display six object snap words, endpoint – perpendicular. These options can be selected when other commands are used, e.g. LINE–SNAPS–endpoint. Note that each line has the format [endpoint]END, i.e. there is no $^\wedge$ C $^\wedge$ C in these lines. Any idea why?

Fifth menu – inserting user-defined blocks

In this menu we will create several blocks and insert them into a new drawing using the screen sub-menu.

1 Open your STDA3 **drawing** file using **FILING–OPENING** from the menu bar – customized menu still displayed?

2 Restore the AutoCAD menu with:
   a) command line **MENU <R>**
   b) activate the R14 folder, e.g. AutoCAD R14 or similar
   c) pick Support from the list in the dialogue box
   d) Menu Template (*.mnu) file type
   e) pick **acad**, then Open and answer Yes to message.

3 The screen should display the full AutoCAD screen menu.

4 Refer to Fig. 11.3 and create the 12 blocks using the information given and your imagination. The block names are given for you and should be used, but the insertion base point is at your discretion.

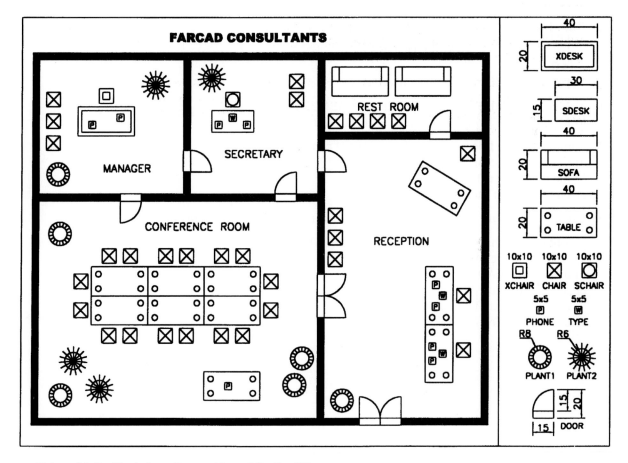

Figure 11.3 Block formation and layout for the fifth menu.

5 Activate Notepad to display the fourth menu and alter as follows:

a) after [THICKS] ^C^C_$S=THICKLINES, enter:

<R>
<R>
<R>
[LAYOUT]^C^C_$S=OFFICE

b) after [DONUT2]^C^C_DONUT 25 30 enter:

<R>
<R>
<R>
<R>

c) after [MAINMENU]^C^C_$S=MAIN enter:

**OFFICE
[FARCAD]
<R>
**[bdesk]^C^C_INSERT XDESK **
**[sdesk]^C^C_INSERT SDESK **
**[bchair]^C^C_INSERT XCHAIR **
**[schair]^C^C_INSERT SCHAIR **
**[chair]^C^C_INSERT CHAIR **
**[rest]^C^C_INSERT SOFA **
**[work]^C^C_INSERT TABLE **
**[lplant]^C^C_INSERT PLANT1 **
**[splant]^C^C_INSERT PLANT2 **
**[ring]^C^C_INSERT PHONE **
**[write]^C^C_INSERT TYPE **
**[in–out]^C^C_INSERT DOOR **
<R>
[MAINMENU]^C^C_$S=MAIN

6 *a*) Menu bar with File–Save to update MYMENU.MNU.

b) Minimize the Notepad screen to return to AutoCAD.

7 Compile the modified menu with the MENU command and:

a) select the r14cust folder

b) file type: Menu Template (*.mnu)

c) pick Mymenu then Open and pick Yes.

8 Use the new menu to design an office layout using Fig. 11.3 as a guide only.

9 Before leaving this exercise 're-load' the AutoCAD menu with:

a) command line MENU <R>

b) folder: AutoCAD R14 (or your equivalent)

c) folder: Support

d) Menu Template File (*.mnu)

e) acad–Open–Yes.

Fifth menu explanation

The new items in this menu are:

a) Three blank lines added to the main screen area after [THICKS] to 'space out' the main menu to 'blank out' the list from the new OFFICE sub-menu.

b) [LAYOUT]^C^C_$S=OFFICE. A new line in the main screen menu. When LAYOUT is selected then cancel any active command and 'goto' the sub-menu labelled OFFICE.

c) Four blank lines in the THICKLINES sub-menu after [DONUT2]. These lines are to 'blank out' the items in the OFFICE sub-menu.

d) **OFFICE: This is the start of the OFFICE sub-menu called with $S=OFFICE.

e) [FARCAD]: the OFFICE sub-menu title followed by a blank <R> line.

f) [bdesk]^C^C_INSERT XDESK. Display the word bdesk in the screen menu area and when selected cancel any active command and activate the AutoCAD INSERT command with the block named XDESK. The \ is a pause for user input, usually the insertion point.

g) The other [???]^C^C_INSERT lines should now be obvious?

h) [MAINMENU]^C^C_$S=MAIN: when selected 'goto' to sub-menu named MAIN, i.e. return to the main menu. This line is preceded by a blank <R> line for spacing purposes.

Sixth menu – user blocks from a pull down menu and an icon menu

In this menu we will create a new drawing using blocks from an icon dialogue box. The blocks were created during the Slide chapter and saved in a drawing called ESTATE. A slide library (ESTATELIB) was also created using the SLIDELIB utility program. A further pull down menu will be added to allow other user defined blocks to be inserted.

1 Open the r14cust\ESTATE drawing from Chapter 9.

2 Check that the five blocks (3BED, 4BED, BUNG, FLAT, SEMI) are 'in' the current drawing using the **?** option of the BLOCK command.

3 Refer to Fig. 11.4 and make four new tree and shrub blocks using:
   a) your imagination
   b) the given block names
   c) the circle centres as the insertion base point.

4 Maximize Notepad from the Windows taskbar to display mymenu.mnu (or open the mymenu.mnu file if you 'closed' Notepad after the previous menu changes).

5 Alter the menu file as follows:
   a) after [perpendicular]PER enter:
   ***POP6
   [GREENERY]
   **[Cherry]^C^C_INSERT TREE1 **
   **[Willow]^C^C_INSERT TREE2 **
   **[Cotoneaster]^C^C_INSERT SHRUB1 **
   **[Berberis]^C^C_INSERT SHRUB2 **
   b) After [THICKS]^C^C_$S=THICKLINES alter as follows:
   [THICKS]^C^C_$S=THICKLINES
   <R>
   [LAYOUT]^C^C_$S=OFFICE
   <R>
   [HOUSES...]^C^C_$I=HOUSEBLK $I=*

c) After [MAINMENU]^C^C_$S=MAIN enter:
***IMAGE
**HOUSEBLK
[House Selection]
**[ESTATELIB(ESTSL1,SEMI)]^C^C_INSERT SEMI **
**[ESTATELIB(ESTSL2,3BED)]^C^C_INSERT 3BED **
**[ESTATELIB(ESTSL3,4BED)]^C^C_INSERT 4BED **
**[ESTATELIB(ESTSL4,FLAT)]^C^C_INSERT FLAT **
[ESTATELIB(ESTSL5,BUNG)]^C^C_INSERT BUNG \.

6   Menu bar with File–Save to update r14cust\mymenu.mnu then minimize Notepad.

7   Recompile the altered menu with:
    a) command line MENU <R>
    b) folder: r14cust
    c) Menu Template file (*.mnu)
    d) mymenu–Open–Yes.

8   Using your menu:
    a) select HOUSES from the screen to display the House Selection dialogue box with your
        house blocks – Fig. 11.5
    b) insert the house blocks – discretion with layout
    c) menu bar with GREENERY to insert the trees/shrubs
    d) optimise your layout
    e) save as r14cust\ESTATE.

9   Select LAYOUT from the screen menu then pick any item and:
    Message: Can't find file in search path, then a list of folders.
    Why is this?

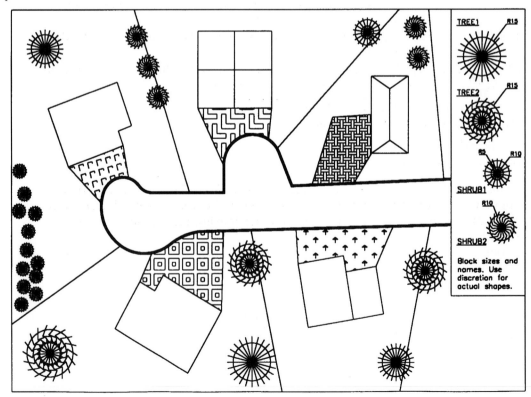

Figure 11.4   Estate layout from menus six and seven.

Sixth menu explanation

Three new sections have been added to the MYMENU.MNU file, these being (*a*), (*b*) and (*c*) in step 5 above.

a) *****POP6**: a new pull down menu. This sequence displays the word GREENERY in the menu bar and allows the user access to the four created tree/shrub blocks/

b) **[THICKS]...**
    <R>
    [LAYOUT]...
    <R>
    [HOUSES]^C^C_$I=HOUSEBLK $I=*

The first four lines 'space-out' the screen menu to allow the inclusion of the new item [HOUSES...] and:

[HOUSES...]:	display HOUSES in the screen menu, the '...' being added to indicate that a dialogue box will follow
^C^C_:	cancel any active command
$I=HOUSEBLK:	'go to' an image (icon) sub-menu called HOUSEBLK. The **$** is the sub-menu 'command' and **I** is used to indicate and image menu.
$I=*:	this is a general AutoCAD command which must be used to display an image tile dialogue box

c) *****IMAGE, etc.**

This sequence will 'load' the house block slides into the dialogue box for user selection and:

***IMAGE:	a section label indicating that an image menu is to follow. This is the same as ***ICON in previous releases
**HOUSEBLK:	the start of the icon sub menu called HOUSEBLK. Thus $I=HOUSEBLK calls up the image sub-menu HOUSEBLK which is displayed with **HOUSEBLK
[House Selection]:	the title of the image dialogue box and is displayed in the title bar

[ESTATELIB(ESTSL1,3BED)]^C^C_INSERT 3BED \: a new line for displaying and inserting slides from a dialogue box, the 'syntax' being important:

ESTATELIB: the name of the slide library file
ESTSL1: the slide name in the slide library file
3BED: the block name

Thus the complete 'line' can be read as:

From the slide library file ESTATELIB, display the slide ESTSL1 in the dialogue box and the block word 3BED in list column. When the slide or the word is selected, cancel any active command and insert the block 3BED into the drawing and wait for user entry for the start point.

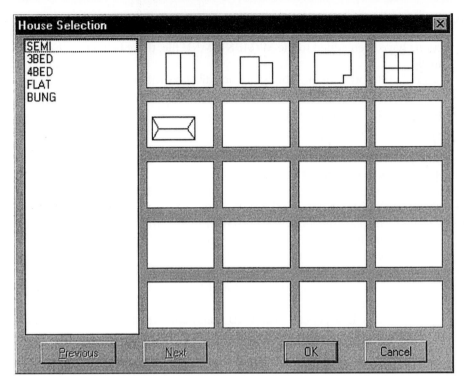

Figure 11.5 House Selection dialogue box.

Seventh menu – inserting created hatch patterns from a dialogue box

This menu is very similar to menu six. The created hatch patterns from Chapter 7 will be 'made available' for selection as icons from a dialogue box. The necessary slides and the slide library file were created in Chapter 9.

1 Still with the ESTATE layout and your menu on the screen?

2 Maximize Notepad from the then:
   a) between [LAYOUT] and [HOUSES...] add:
      [HATCH...]^C^C_$I=HATPATS $I=*
   b) after the last [ESTATELIB] enter:
      **HATPATS
      [My Own Created Hatch Patterns]
      **[HATCHLIB(HATSL1,HANG)]^C^C_–BHATCH P HANG 0.5 15 **
      **[HATCHLIB(HATSL2,DBOX)]^C^C_–BHATCH P DBOX 1 0 **
      **[HATCHLIB(HATSL3,WEAVE)]^C^C_–BHATCH P WEAVE 1 0 **
      **[HATCHLIB(HATSL4,ELLS)]^C^C_–BHATCH P ELLS 0.5 0 **
      **[HATCHLIB(HATSL5,ARROWS)]^C^C_–BHATCH P ARROWS 0.25.0 **

3 *a*) File–Save to update r14cust\mymenu.mnu
   b) minimize the Notepad screen to return to AutoCAD.

4 Compile your modified menu with:
   a) command line MENU
   b) Menu Template file (*.mnu) – r14cust folder
   c) pick Mymenu–Open–Yes.

5 The screen menu should display HATCH...

6 Select HATCH... from the screen menu and:
   a) Hatch pattern dialogue box displayed – Fig. 11.6
   b) select any pattern (name or icon) and add hatching to the estate layout as Fig. 11.4, then save the drawing.

7 *Note.*
   Using the hatch patterns from the dialogue box allows the user to select an 'internal point' for hatching, but there is no preview option.

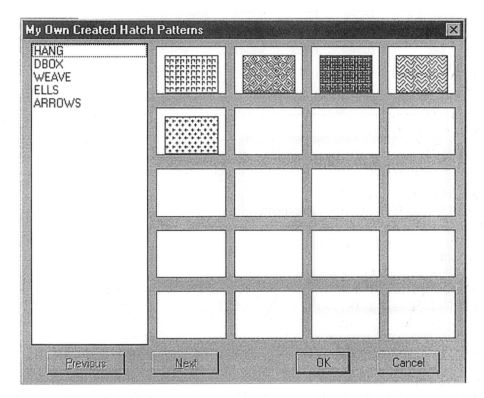

Figure 11.6 Hatch icon dialogue box.

Seventh menu explanation

The new items add to the MYMENU.MNU file are:

a) [HATCH...]^C^C_$I=HATPATS $I=*
When HATCH... is selected from the screen menu, cancel any active command and 'go to' the image sub-menu named HATPATS

b) **HATPATS
The start of the image sub-menu called with $I=HATPATS. Note that this sub-menu is still 'within' the ***IMAGE section label used to activate the house block sub-menu

c) [My Own Created Hatch Patterns]
The name of the hatch pattern dialogue box

d) [HATCHLIB(HATSL1,HANG)]^C^C_-BATCH P HANG 0.5 15 \
HATCHLIB: the name of the slide library file containing the slides
HATSL1,HANG: display the slide HATSL1 as an icon in the dialogue box and display the word HANG in the list column

^C^C_:	cancel any active command
-BHATCH:	the command line boundary hatch command to 'bypass' the boundary hatch dialogue box
P:	the properties option of the BHATCH command which allows the hatch pattern name to be entered
HANG:	the name of the hatch pattern to be used
0.5:	the hatch pattern scale
15:	the hatch pattern angle
\:	a pause for user entry.

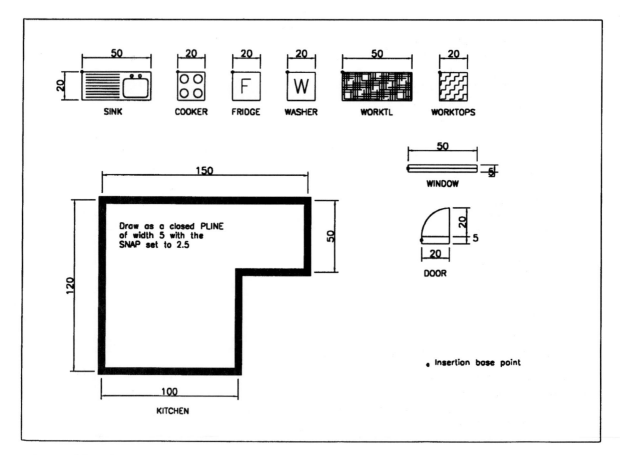

Figure 11.7 Block information for drawing KITCHEN for use with the tablet menu.

Eighth menu – a tablet menu

This menu can only be used if you use (or have access) to a digitizing tablet compatible with AutoCAD. The menu requires a puck as a pointing device and will not work with the mouse. If you do not have access to a tablet then forget this exercise and the digitizing exercise which follows. Proceed to the toolbars customization exercise. The tablet menu will insert created blocks by 'icon' selection.

1 Open your STDA3 drawing file and activate the AutoCAD R14 menu.

2 Refer to Fig. 11.7 and:
  a) draw the nine objects using the information given
  b) make blocks of each object with:
    i) the block name given
    ii) the top-left corner of each object as the insertion base point – suggestion only
    iii) do not add the dimensions.

3 At this point save the drawing as r14cust\KITCHEN.

4 Maximize Notepad from the Windows taskbar and:
  a) After [THICKS] enter:
  [TABLET]^C^C_$S=TABLMEN   line1
  b) Before ***IMAGE enter:
  **TABLMEN   line2
  <R> 16 times   line3
  [MAINMENU]^C^C_$S=MAIN   line4
  ***TABLET1   line5
  **^C^C_INSERT SINK **   line6
  **^C^C_INSERT COOKER **
  **^C^C_INSERT FRIDGE **
  **^C^C_INSERT WASHER **
  **^C^C_INSERT WORKTL **
  **^C^C_INSERT WORKTS **
  **^C^C_INSERT DOOR **
  **^C^C_INSERT WINDOW **
  **^C^C_INSERT KITCHEN **
  ***TABLET2
  ^C^C_COPY
  ^C^C_MOVE
  ^C^C_ROTATE
  ^C^C_MIRROR
  ^C^C_LINE
  ^C^C_DTEXT
  ^C^C_ERASE
  'REDRAW
  'REDRAW.

5 *a*) File–Save to update MYMENU.MNU
  b) minimize Notepad screen to return to AutoCAD.

6 Compile the modified menu with:
  a) command line MENU
  b) menu template file (*.mnu) – r14cust folder
  c) pick MYMENU–Open–Yes.

Setting the tablet

The tablet is a pointing device and has to be 'installed' to the system requirements. This is achieved with the following sequence:

1 At the command line enter **PREFERENCES <R>** and:
   prompt   *Preferences dialogue box*
   respond   pick the Pointer tab
   prompt   *list of pointing devices*
   respond   1. scroll and pick your tablet, e.g. Summagraphics MM
               2. pick Set Current
   prompt   *Supported model list*
   respond   enter as required, e.g. 2
   prompt   *number of buttons*
   respond   enter as required, e.g. 4
   prompt   *Serial port for digitizer*
   respond   enter as required, e.g. COM1 or COM2
   prompt   *Pointer tab screen returned*
   respond   1. pick Digitizer and mouse
               2. pick Apply.

2 The drawing screen will be returned and both the mouse and puck should be active and available to the user.

Configuring the table

1 Cut-out the tablet overlay (Fig. 11.8) and fix it to your tablet. The overlay consists of a screen drawing area and two menu areas, one for the kitchen blocks and the other for commands. The two menu areas consist of one column and nine rows. The order of the symbols and commands in each menu area is the same as the order in the menu file. This is **essential** with tablet menus.

2 At the command line enter **TABLET <R>** and:
   prompt   *Option(ON/OFF/CAL/CFG)*
   enter   **CFG <R>** – the configure option
   prompt   *Enter number of tablet menus desired(0-4)*
   enter   **2<R>**
   prompt   *Digitize upper left corner of menu area 1*
   respond   **pick pt1**
   prompt   *Digitize lower left corner of menu area 1*
   respond   **pick pt2**
   prompt   *Digitize lower right corner of menu area 1*
   respond   **pick pt3**
   prompt   *Enter the number of columns for menu area 1*
   enter   **1 <R>**
   prompt   *Enter the number of rows for menu area 1*
   enter   **9 <R>**
   prompt   *Digitize upper left corner of menu area 2*
   respond   **pick pt4**
   prompt   *Digitize lower left corner of menu area 2*
   respond   **pick pt5**
   prompt   *Digitize lower right corner of menu area 2*
   respond   **pick pt6**
   prompt   *Enter the number of columns for menu area 2*
   enter   **1 <R>**

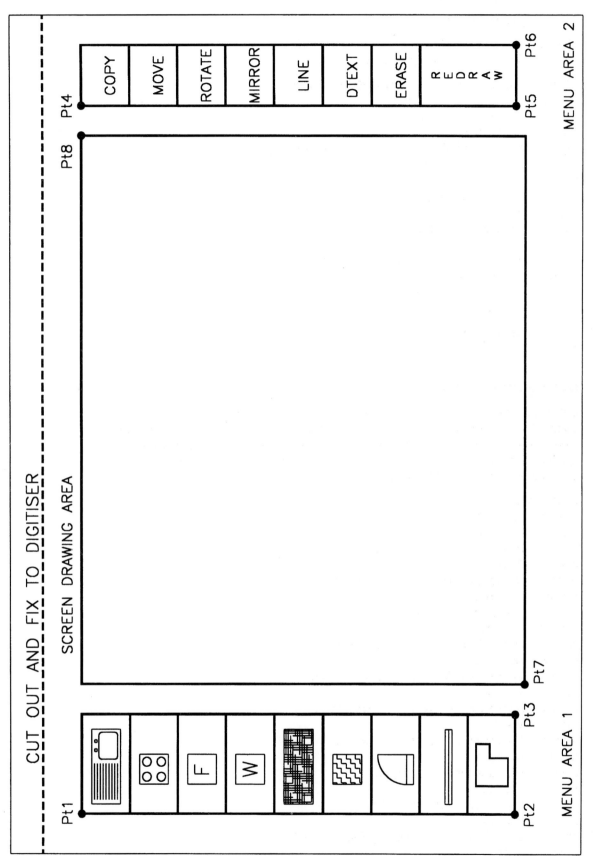

Figure 11.8 Overlay for use with tablet menu.

prompt	Enter the number of rows for menu area 2
enter	**9 <R>**
prompt	Do you want to respecify the Fixed Screen pointing area?
enter	**Y <R>**
prompt	Digitize lower left corner of Fixed Screen pointing area
respond	**pick pt7**
prompt	Digitize upper right corner of Fixed Screen pointing area
respond	**pick pt8**
prompt	Do you want to specify the Floating Screen pointing area?
enter	**N <R>**

Using the modified menu

1 Modified menu recompiled?

2 Open the saved drawing KITCHEN or erase all objects from the screen.

3 Select from the screen TABLET and the screen menu area will only display MAINMENU, which will return your full menu.

4 Refer to Fig. 11.9 and use the tablet menu to construct some kitchen layouts of your own design. I set the snap to 2.5 and the grid to 5 for this exercise – why?

5 When complete, save as KITCHEN in your r14cust folder.

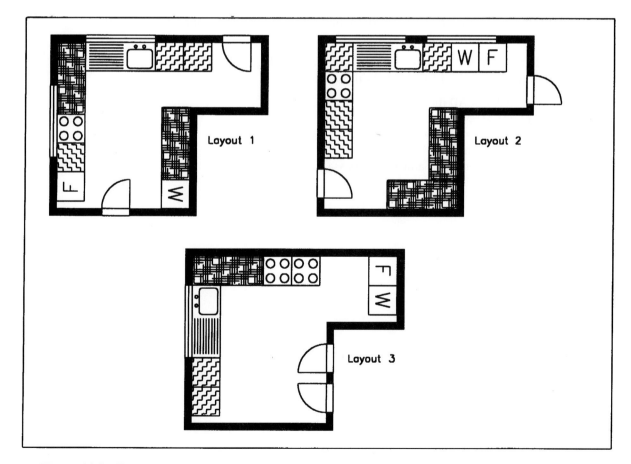

Figure 11.9 Three kitchen layouts using the tablet menu.

Eighth menu explanation

The new items added to the menu are:

line1:   [TABLET]^C^C_$S=TABLMEN
         Displays the word TABLET in the screen menu area and when selected, activate
         the sub-menu called TABLMEN

line2:   ****TABLMEN**
         The start of the sub-menu, called with $S=TABLMEN

line3:   **<R> 16 times**
         Used to 'blank-out' the screen menu item names

line4:   **[MAINMENU]^C^C_$S=MAIN**
         The only item displayed in the screen menu area and allows the main menu
         to be 'restored'

line5:   *****TABLET1**
         The start of tablet menu area 1. ***TABLETn is a main section label.

line6:   **^C^C_INSERT SINK **

         Insert the created block SINK which is the first item in menu area 1. The order
         of the menu blocks **must be the same** as the order of the order on the
         template overlay.

Notes

1   The rest of the new items should be easy to understand?

2   The items mentioned in lines 1–4 are not really necessary in this menu – think about
    this!

3   The complete MYMENU.MNU at this stage is displayed in Fig. 11.10. This menu has been
    imported into the standard STDA3 drawing sheet.

4   Restore the full AutoCAD menu with:
    a) command line MENU
    b) the AutoCAD R14 – Support folder
    c) file type: Menu Template (*.mnu)
    d) pick acad.mnu, then Open and Yes to message.

```
***AUX1                          [MAINMENU]^C^C_$S=MAIN              ***POP2
;                                ***TABLET1                          [MODIFYING]
***SCREEN                        ^C^C_INSERT SINK  \                 [ERASEWIN]^C^C_ERASE W
**MAIN                           ^C^C_INSERT COOKER  \               [ERASELAST]^C^C_ERASE L
[MYMENU]                         ^C^C_INSERT FRIDGE  \               [COPY]^C^C_COPY
                                 ^C^C_INSERT WASHER  \               [MIRROR]^C^C_MIRROR
[STRAITS]^C^C_LINE               ^C^C_INSERT WORKTL  \               [ROTATE]^C^C_ROTATE
[ROUNDS]^C^C_CIRCLE              ^C^C_INSERT WORKTS  \               [MOVE]^C^C_MOVE
[WORDS]^C^C_DTEXT                ^C^C_INSERT DOOR  \                 [->ARRAYS]
                                 ^C^C_INSERT WINDOW  \               [RECTANGULAR]^C^C_ARRAY  \  R
[RUBOUT]^C^C_ERASE               ^C^C_INSERT KITCHEN  \              [<-POLAR]^C^C_ARRAY  \  P
                                 ***TABLET2                          ***POP3
[ZOOMIN]^C^C_ZOOM W              ^C^C_COPY                           [FILING]
[ZOOMOUT]^C^C_ZOOM P             ^C^C_MOVE                           [SAVING]^C^C_SAVE
                                 ^C^C_ROTATE                         [OPENING]^C^C_OPEN
[THICKS]^C^C_$S=THICKLINES       ^C^C_MIRROR                         [NEWING]^C^C_NEW
[TABLET]^C^C_$S=TABLMEN          ^C^C_LINE                           [ENDING]^C^C_QUIT
[LAYOUT]^C^C_$S=OFFICE           ^C^C_DTEXT                          ***POP4
[HATCH...]^C^C_$I=HATPATS $I=*   ^C^C_ERASE                          [BOXES]
[HOUSES...]^C^C_$I=HOUSEBLK $I=* 'REDRAW                             [layers]^C^C_DDLMODES
**THICKLINES                     'REDRAW                             [aids]^C^C_DDRMODES
[POLYLINE]                       ***IMAGE                            [units]^C^C_DDUNITS
                                 **HOUSEBLK                          [change]^C^C_DDCHPROP
[PLINE1]^C^C_PLINE  \W 5 5       [House Selection]                   ***POP5
[PLINE2]^C^C_PLINE  \W 10 0      [ESTATELIB(ESTSL1,SEMI)]^C^C_INSERT SEMI  \   [SNAPS]
[DONUT1]^C^C_DONUT 0 10          [ESTATELIB(ESTSL2,3BED)]^C^C_INSERT 3BED  \   [endpoint]END
[DONUT2]^C^C_DONUT 25 30         [ESTATELIB(ESTSL3,4BED)]^C^C_INSERT 4BED  \   [midpoint]MID
                                 [ESTATELIB(ESTSL4,FLAT)]^C^C_INSERT FLAT  \   [centre]CEN
                                 [ESTATELIB(ESTSL5,BUNG)]^C^C_INSERT BUNG  \   [intersection]INT
                                 **HATPATS                           [quadrant]QUA
                                 [My Own Created Hatch Patterns]     [perpendicular]PER
                                 [HATCHLIB(HATSL1,HANG)]^C^C_-BHATCH P HANG 0.5 15  \   ***POP6
                                 [HATCHLIB(HATSL2,DBOX)]^C^C_-BHATCH P DBOX 1 0  \     [GREENERY]
                                 [HATCHLIB(HATSL3,WEAVE)]^C^C_-BHATCH P WEAVE 1 0  \   [Cherry]^C^C_INSERT TREE1  \
                                 [HATCHLIB(HATSL4,ELLS)]^C^C_-BHATCH P ELLS 0.5 0  \   [Willow]^C^C_INSERT TREE2  \
[MAINMENU]^C^C_$S=MAIN           [HATCHLIB(HATSL5,ARROWS)]^C^C_-BHATCH P ARROWS 0.25 0  \  [Cotoneaster]^C^C_INSERT SHRUB1  \
**OFFICE                         ***POP1                             [Berberis]^C^C_INSERT SHRUB2  \
[FARCAD]                         [DRAWING]
                                 [ARC]^C^C_ARC C
[bdesk]^C^C_INSERT XDESK  \      [DONUTUSER]^C^C_DONUT
[sdesk]^C^C_INSERT SDESK  \      [ELLIPSE]^C^C_ELLIPSE C
[bchair]^C^C_INSERT XCHAIR  \    [PLINEUSER]^C^C_PLINE  \W
[schair]^C^C_INSERT SCHAIR  \    [->CIRCLES]
[chair]^C^C_INSERT CHAIR  \      [CENRAD]^C^C_CIRCLE
[rest]^C^C_INSERT SOFA  \        [CENDIA]^C^C_CIRCLE  \D
[work]^C^C_INSERT TABLE  \       [THREE]^C^C_CIRCLE 3P
[lplant]^C^C_INSERT PLANT1  \    [<-TTRAD]^C^C_CIRCLE TTR
[splant]^C^C_INSERT PLANT2  \    [->MULTIS]
[ring]^C^C_INSERT PHONE  \       [PENTAGON]^C^C_POLYGON 5  \
[write]^C^C_INSERT TYPE  \       [HEXAGON]^C^C_POLYGON 6  \
[in-out]^C^C_INSERT DOOR  \      [OCTAGON]^C^C_POLYGON 8  \
                                 [<-DECAGON]^C^C_POLYGON 10  \
[MAINMENU]^C^C_$S=MAIN
**TABLMEN
```

Figure 11.10   Complete listing of C:\r14cust\MYMENU.MNU.

Digitizing a drawing

While this is not a menu creation topic, it is still an interesting exercise as it involves digitizing a drawing from a tablet overlay. The tablet is calibrated to the overlay and the overlay then used to complete the drawing. To demonstrate how this is achieved:

1   Open your STDA3 template file and display toolbars as required.

2   Cut-out Fig. 11.11 and fix it to your digitizing tablet.

3   At the command line enter **TABLET <R>** and:
   prompt   Option (ON/OFF/CAL/CFG)
   enter   **CAL <R>** – the calibrate option
   prompt   Digitize point #1
   respond   with the puck, **pick pt1**
   prompt   Enter coordinates for point #1
   enter   **60,50 <R>**
   prompt   Digitize point #2
   respond   with the puck, **pick pt2**
   prompt   Enter coordinates for point #2
   enter   **310,50 <R>**
   prompt   Digitize point #3 (or press ENTER to end)
   respond   **<RETURN/ENTER>**.

4   The status bar should now display the word **Tablet** indicating that the tablet is 'active'.

5   Try and pick the LINE icon from the Draw toolbar – not possible?

6   To activate the toolbars when the tablet is being used, it is necessary to 'toggle' between the tablet and the toolbars/menu bar items/screen menu (if displayed). This is achieved with the **F4** function key, i.e. if you want to select the LINE icon:
   a) press F4 to 'activate' the toolbars/menu bar
   b) pick the required icon/menu bar item
   c) press F4 to 'return to the tablet'
   d) repeat the F4 toggle for other commands
   e) Note: command line entry is possible at all times.

7   Using the overlay, complete the drawing by digitizing. The more difficult items to Digitize are:
   a) circles – snap to centre points
   b) donuts – the ID and OD
   c) the polylines
   d) the hatching.

8   Save the completed drawing if required, although we will not refer to it again.

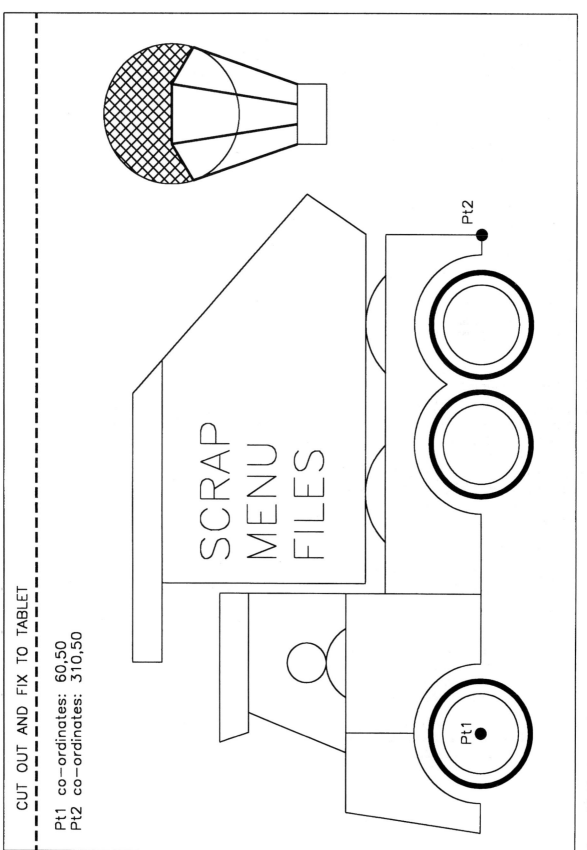

CUT OUT AND FIX TO TABLET

Pt1 co-ordinates: 60,50
Pt2 co-ordinates: 310,50

SCRAP
MENU
FILES

Pt2

Pt1

Figure 11.11 Drawing for digitizing with the tablet.

Toolbar customization

Release 14 users are probably more than content with the existing dialogue boxes, but it is possible to alter the existing icons in a toolbar or add other icons if required. In this section we will investigate how a new toolbar can be created with icons of our choice, so:

1 Open your STDA3 template file with the AutoCAD R14 menu acad.mnu.

2 Refer to Fig. 11.12 and select from the menu bar **View–Toolbars** and:
   prompt   *Toolbars dialogue box* – with list of toolbars
   respond **pick New**
   prompt   *New Toolbar dialogue box*
   respond 1. Toolbar name: enter **MINE**
              2. Menu Group: ACAD – usually is this
              3. pick OK
   prompt   *Toolbars dialogue box*
   with   1. MINE listed and active (*X*)
              2. new toolbar displayed at top-centre of screen as Fig. 11.12(a)
   respond pick Close.

3 Move the new toolbar to the left-centre of the screen.

(a)New toolbar added

(b)ARRAY icons added

(c)DIVIDE and MEASURE icons added

(d)Custom flyout icon added

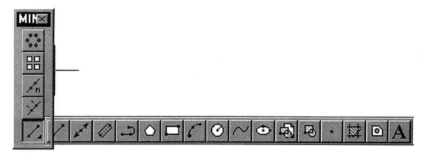
(e)The flyout from the LINE icon

Figure 11.12 Toolbar customization.

4 Repeat the View–Toolbars menu bar selection and:

   prompt   *Toolbars dialogue box*

   respond   **pick Customize**

   prompt   *Customize dialogue box* – with list of Categories

   respond   1. scroll at Categories

              2. pick Modify

   prompt   *Customize Toolbar dialogue box of Modify icons*

   respond   1. pick the Polar Array icon and hold down the left button

              2. drag the icon into the new MINE toolbar

              3. release the left button

              4. pick and drag the Rectangular Array icon into the new toolbar – Fig. 11.12(b)

              5. scroll at Categories and pick Draw

              6. pick and drag the Divide and Measure icons into the new toolbar – Fig. 11.12(c)

              7. pick Close and Close.

5 At the command line enter **TOOLBAR <R>** and:

   prompt   *Toolbar dialogue box*

   respond   pick Customize

   prompt   *Customize Toolbar dialogue box*

   respond   1. scroll and pick Custom Flyout

              2. pick and drag the Line Flyout icon (with black triangle) into the new toolbar – Fig. 11.12(d)

              3. pick Close and Close.

6 Move the cursor to the line icon of the new toolbar and **double-right click** and:

   prompt   *Toolbars dialogue box*

   then   *Flyout Properties dialogue box*

   with   Line as Flyout name

   respond   1. pick ACAD.Draw

              2. pick Show This Button's Icon

              3. pick Apply

              4. Cancel the Flyout Properties dialogue box (top left)

              5. pick Close.

7 When the LINE icon is selected from the new toolbar, the draw flyout menu will be displayed as Fig. 11.12(e). This will allow the user additional icon selection.

8 The new toolbar can be cancelled/activated from the Toolbar dialogue box in the normal manner.

9 This completes the toolbar customization section.

Summary

1 Customized menu files are text files written by the user. They have the extension **.MNU**. All user-written menus are called TEMPLATE menus and are compiled by AutoCAD.

2 AutoCAD has several menu 'types' in the AutoCAD R14/SUPPORT folder. These menus are:
acad.mnu: the main template menu
acad.mnc: the compiled menu
acad.mnr: the resource menu
acad.mns: the source menu.

3 Template menus are identified with section labels as follows:

section label	menu area
***SCREEN	screen menu area at right of monitor
***POPn	pull-down menus (0–16)
***IMAGE	image tile menus(icon menus)
***AUXn	auxiliary menus, e.g. buttons (1–4)
***TABLETn	tablet menus (1–4).

4 Menu syntax is determined by special characters, and the most common items are listed in the following table:

character	usage
***	main section label
**	sub-menu label
[]	brackets used to display items in the screen menu, pull-down menus and icon menus
;	equivalent to a <RETURN> in a menu item line
$	loads a menu section, e.g. a sub-menu
^C^C	cancels any active command
\	pause for user entry.

5 Slides can be included in image tile menus, but a slide library containing the slides should be created.

6 Tablet menus are generally use with user-defined blocks.

7 A tablet can be used to digitize a drawing. This is not possible with the mouse pointing device.

8 Toolbars can be customized to user requirements.

9 The menu bar selection **Tools–Customize Menus** allows:
   a) menus to be loaded/unloaded as required
   b) menu bar items to be inserted/renamed.

Assignment

Two activities have been included, one which requires modification to the menu written in the chapter while the other requires a drawing to be digitized.

Activity 8: Modified menu with user-defined blocks

1 Open your STDA3 template file and refer to Fig. 11.13.

2 Draw the ten symbols with the sizes given using your discretion for sizes not given.

3 Add attributes to the RES, NPN, PNP, DIO, CAP and INC blocks, the attribute information being:
Tag: V
Prompt: Enter value
Default: aa.

4 Create blocks of the ten symbols using the block names and insertion points given.

5 Modify the MYMENU.MNU file to allow the 10 blocks to be inserted into a drawing from a pull-down menu.

6 Add a cascade effect to the pull-down menu which will allow attributes to be extracted in either CDF or SDF format. Figure 11.14 displays the pull-down menu idea.

7 Recompile the MYMENU.MNU template menu file and use the menu to complete the electrical circuit, adding the component information as attribute data in response to the prompts.

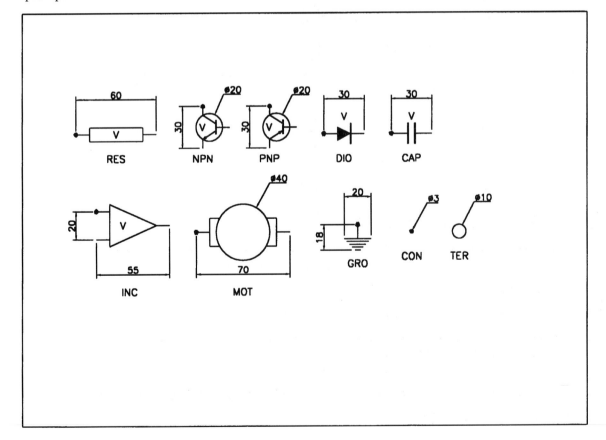

Figure 11.13 Block information for menu activity 8 with tags in position.

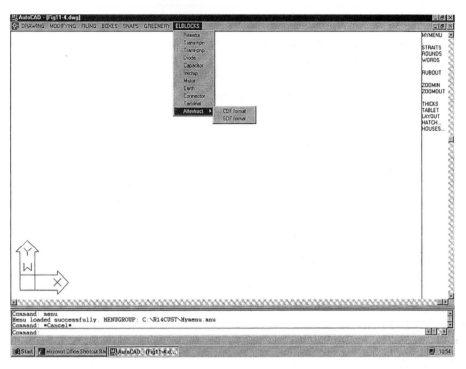

Figure 11.14 Cascade menu effect.

Note: The following may help with the drawing:
   a) limits: 0,0 to 1000,600 then zoom–all
   b) snap: 5
   c) MIRRTEXT: 0 – any idea why?

8 Using Notepad, write a template file named ELECT.TXT, consisting of two lines:
   BL:NAME   C005000
   V   C005000.

9 Extract the attribute data in both CDF and SDF formats, the extract file names being:
   a) CDF format: r14cust\ELECTCDF.TXT
   b) SDF format: r14cust\ELECTSDF.TXT.

10 Save the completed drawing as **r14cust\ELECTLAY**.

Activity 9: A digitized drawing

This activity is slightly different from normal. I have included the first ever drawing I managed to digitize and it requires a bit of work from the user.

1 Open your STDA3 template file.

2 Cut out the Activity 8 drawing and fix it to your tablet.

3 Calibrate the tablet with the TABLET–CAL keyboard entry with:
   a) digitize point #1: pt1 with coordinates 0,0
   b) digitize point #2: pt2 with coordinates 210,290.

4 Digitize the outline with the LINE command and add other lines and polylines to suit.

5 The poem can be added if required.

6 Save the completed drawing.

Chapter 12

Data exchange

Transferring data from CAD is always one of the justifications for the expense of installing a CAD system. This transfer may be to other CAD systems, for CNC machining, database transactions, spreadsheet usage, etc. In an earlier chapter we created attribute extract files and extracted **TEXT** information in CDF and SDF formats. In this chapter we will investigate several methods of transferring **DRAWING** information in/out of AutoCAD.

AutoCAD Release 14 supports both graphical vector and non-graphic raster file formats including:

a) Vector format:

3D Studio	3DS (extension)
Drawing Interchange	DXF
PostScript	EPS
Windows Metafile	WMF

b) Solid model format

ACIS Solid Object	SAT
Solid Stereo-lithography	STL

c) Raster format:

Bitmap	BMP
True vision targa	TGA
Paintbrush	PCX
Tagged Image	TIFF.

The drawings

1   Open your r14cust\STDA3 **drawing** file and refer to Fig. 12.1(a).

2   Draw the component as shown using layers correctly. Add the text and dimensions. Position the circle centres at the point 100,120.

3   Erase the border then save the drawing as **r14cust\COUPLING**.

4   Erase all objects from the screen and make a new layer MODEL, colour red and current. Refer to Fig. 12.1(b).

5   Purge layers CL, DIM, HID, OUT, SECT and TEXT.

6   With the menu bar sequence **View–3D Viewpoint–SE Isometric**, change the viewpoint.

7   Select from the menu bar **Draw–Solids–Box** and enter:
    a) corner: **0,0,0 <R>**
    b) options: enter **C <R>** – cube option
    c) length: **200 <R>**
    d) colour: red.

8   Zoom–all.

9   Menu bar with **Draw–Solids–Box** and create a solid box with:
    a) corner: **200,200,200 <R>**
    b) options: enter **L <R>** – length option
    c) length: **–100**, width: **–120**, height: **–80**
    d) colour: blue.

10 Create a cylinder with **Draw–Solids–Cylinder** and:
    a) centre: **100,100,0 <R>**
    b) radius: **50 <R>**
    c) height: **200 <R>**
    d) colour: green.

11 At the command line enter **ISOLINES <R>** and set to **24**, then REGEN the drawing – better cylinder 'definition'?

12 Menu bar with **Modify–Boolean–Subtract** and:
    a) pick the red box then right-click
    b) pick the blue box and green cylinder then right-click.

13 At command line enter **ZOOM <R>** and:
    a) options: enter **C <R>**
    b) centre point: enter **100,100,100 <R>**
    c) magnification: enter **400 <R>**.

14 Save the model as **r14cust\SOLCOMP**.

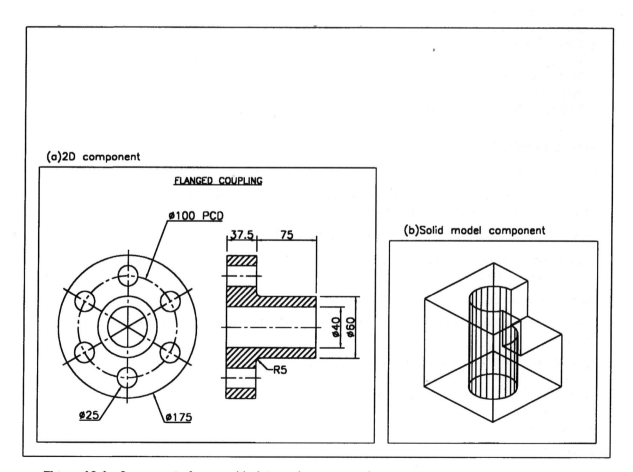

Figure 12.1 Components for use with data exchange examples.

Exporting data – graphic vectors
Creating a DXF export file

Drawing Interchange Format (DXF) files can be read by other CAD systems and most CNC systems.

1 Open the r14cust\COUPLING drawing file.

2 Menu bar with **File–Export** and:
   prompt   Export Data dialogue box
   respond   1. ensure r14cust current, i.e. Save in
            2. scroll and pick file type **AutoCAD R14 DXF (*.dxf)**
            3. File name: COUPLING.dxf
            4. pick Save.

3 At the command line enter **DXFOUT <R>** and:
   prompt   Create DXF File dialogue box
   with   COUPLING listed
   respond **pick Options**
   prompt   Export Options dialogue box
   respond   1. ASCII format current
            2. Decimal places of accuracy: 6
            3. pick Select Objects (tick in box)
            4. pick OK
   prompt   Create DXF File dialogue box
   respond   1. Alter file name: COUPL_1
            2. pick Save
   prompt   Select objects
   respond   window the left view then right-click.

4 Repeat the DXFOUT command and:
   a) alter file name to COUPL_2
   b) pick Options
   c) select BINARY format
   d) cancel the Select Objects
   e) pick OK
   f) pick Save from Create DXF File dialogue box.

5 We have created three DXF output files:
   a) COUPLING: the complete component in ASCII format
   b) COUPL_1: the left view in ASCII format
   c) COUPL_2: the complete component in BINARY format.

Creating a WMF export file

Window Metafile Format (WMF) files are good representations of drawings as they contain both screen vectors and raster graphics.

1 Open the drawing file r14cust\COUPLING.

2 Menu bar with **File–Export** and:
   prompt   Export Data dialogue box
   respond   1. ensure r14cust current
            2. File type: Metafile (*.wmf)
            3. File name: COUPLING
            4. pick Save
   prompt   Select objects
   respond   window the complete component then right-click.

3 The command entry is **WMFOUT**.

Creating a PostScript export file

PostScript files are graphical files particularly suited to electronic publishing work. PostScript fonts can be used in AutoCAD to enhance the AutoCAD text fonts. PostScript files have the extension .EPS which is an abbreviation for encapsulated PostScript.

1 Open the r14cust\COUPLING drawing.

2 Menu bar with **File–Export** and:
   prompt   *Export Data dialogue box*
   respond   1. r14cust current
              2. scroll and pick file type: Encapsulated PS (*.eps)
              3. file name: COUPLING.eps
              4. pick Save.

3 The command line entry is **PSOUT**.

Exporting data – solid models

Creating an ACIS export file

1 Open the solid model r14cust\SOLCOMP.

2 Menu bar with **File–Export** and:
   prompt   *Export Data dialogue box*
   respond   1. r14cust current
              2. scroll and pick file type ACIS (*.sat)
              3. file name: SOLCOMP.sat
              4. pick Save
   prompt   *Select objects*
   respond   pick the model the right-click.

3 The command line entry is **ACISOUT**.

Creating a Stereo-lithography export file

1 Open drawing r14cust\SOLCOMP.

2 Menu bar with **File–Export** and:
   prompt   *Export Data dialogue box*
   respond   1. r14cust current
              2. file type: Lithography (*.stl)
              3. file name: SOLCOMP.stl.

3 Pick Save
   prompt   *Select objects*
   respond   pick the model then right-click.

4 Command line entry is **STLOUT**.

Exporting data – raster screen images

Raster files create an image of the drawing screen which can be used in software packages where accuracy is not of the greatest importance, e.g. in animation, logo design, etc. Commonly used raster formats are:

a) bitmap (BMP) files
b) paintbrush (PCX) files
c) true vision targa (TGA) files
d) tagged image (TIFF) files.

Creating a BMP export file

1 Open the r14cust\COUPLING drawing file.

2 Menu bar with **File–Export** and:
   prompt   Export Data dialogue box
   respond   1. r14cust current
            2. file type: Bitmap (*.bmp)
            3. file name: COUPLING.bmp
            4. pick Save
   prompt   Select objects
   respond   window the component then right-click.

3 The command line entry is **BMPOUT**.
   Only the bitmap export file has been considered in this chapter. The other raster file formats are really for rendered images which will not be discussed in this book.

Checking the created files

1 Activate the MS-DOS (Command) prompt from the Windows taskbar and:
   prompt   C:|> (or similar)
   enter   **cd\r14cust <R>**
   then   **dir COUPLING.* <R>**
   prompt   file listing with:

size	file name
42194	COUPLING.dxf
35260	COUPLING.dwg
9652	COUPLING.wmf
496630	COUPLING.bmp
12878	COUPLING.eps.

2 Study the list and note that the various file formats use different amounts of memory. The BMP file is the largest. The DXF format 'takes' more memory than the original DWG file. Any idea why this is?

3 Minimize or cancel the command prompt screen to return to AutoCAD.

Importing files into AutoCAD

Now that several different file formats have been created, we will investigate how these files can be imported into AutoCAD. This may seem to be a bit stupid as we have the original drawings, but it will be interesting to 'see' if the imported file types display the drawing correctly.

Importing a DXF file

1 Menu bar with **File–New** and:
  a) pick Start from Scratch
  b) pick Metric then OK.

2 At the command line enter **DXFIN <R>** and:
  prompt *Select DXF File dialogue box*
  respond 1. select the r14cust folder name
        2. pick COUPLING then OK.

3 The exported DXF file of the coupling will be displayed with text, dimensions and hatching.

4 *Questions.*
  a) circle centres at 100,120?
  b) layers imported correctly?
  c) objects can be modified as expected?
  d) complete drawing has been imported correctly?

5 Try and import the other DXF files, i.e. COUPL_1 and COUPL_2 – not possible? Read the prompt and try what is suggested.

6 Menu bar with File–New as step 1 and import the other DXF files.

7 Do not save any of these imported files.

Importing a WMF file

1 Menu bar with **File–New** and:
  a) pick Start from Scratch
  b) pick Metric then OK.

2 Menu bar with **Insert–Windows Metafile** and:
  prompt *Import WMF dialogue box*
  respond 1. select r14cust
        2. pick COUPLING – preview obtained
        3. pick OK
  prompt *Insertion point* and note ghost image and drag effect
  enter **0,290 <R>**
  prompt *X scale.* and enter: **1 <R>**
  prompt *Y scale.* and enter: **1 <R>**
  prompt *Rotation.* and enter: **0 <R>**.

3 The coupling 'drawing' will be displayed as Fig. 12.2(a).

4 *Questions.*
  a) why not 0,0 as the insertion point?
  b) is the display a 'drawing' or a 'block'?
  c) can it be exploded?
  d) what about layers?
  e) why the 'frame'?

5 At the command line enter **WMFOPTS <R>** and:
  prompt *Import options*
  respond cancel both the Wire Frame and Wide Lines (no tick in the boxes) then pick OK.

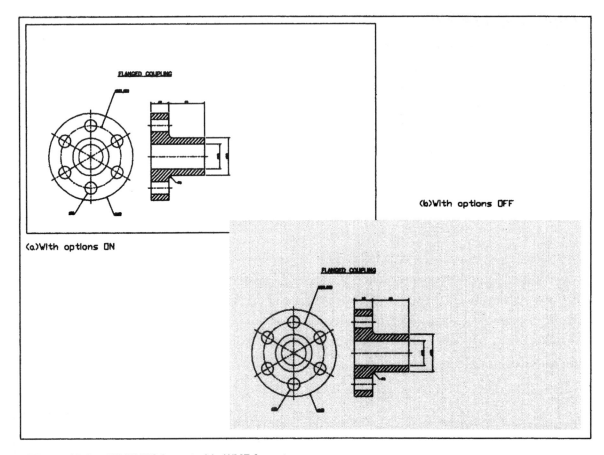

Figure 12.2 COUPLING imported in WMF format

6 At the command line enter **WMFIN <R>** and:
   prompt   *Import WMF dialogue box*
   respond   1. select r14cust
                2. pick COUPLING then OK
   prompt   *Insertion point* and enter: **150,150 <R>**
   prompt   *X scale.* and enter: **1 <R>**
   prompt   *Y scale.* and enter: **1 <R>**
   prompt   *Rotation.* and enter: **0 <R>**.

7 The imported file is displayed as Fig. 12.2(b).

8 *Task.*
   Is it possible to import the COUPLING.WMF file so that the circle centres are positioned at the point 100,120? An insertion point and *X* and *Y* scale values are required.

Importing an EPS file

1 Menu bar with File–New and Start from Scratch–Metric–OK.

2 At the command line enter **PSDRAG <R>** and:
   prompt   *PSIN drag mode <0>* and enter: **1 <R>**.

3 At the command line enter **PSQUALITY <R>** and:
   prompt   *New value for PSQUALITY<??>* and enter: **15 <R>**.

4 Menu bar with **Insert–Encapsulated PostScript** and:
   prompt   *Select PostScript File dialogue box*
   respond 1. pick r14cust
               2. pick COUPLING then Open
   prompt   *Insertion point <0,0,0>* and enter: **0,0 <R>**
   and   ghost image of file
   prompt   *Scale factor* and enter: **100 <R>**.

5 The COUPLING.EPS file is displayed as Fig. 12.3(a) and is not very impressive?

6 At the command line enter PSQUALITY and alter the value to 50.

7 At the command line enter **PSIN <R>** and:
   a) select the r14cust\COUPLING file
   b) pick Open
   c) Insertion point: 0,90
   d) Scale factor: 100
   e) Fig. 12.3(b).

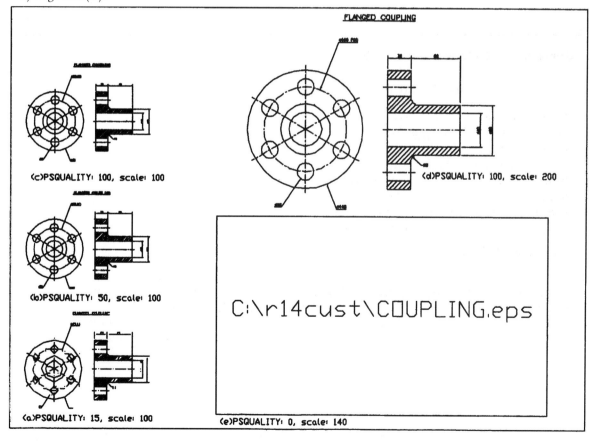

Figure 12.3   COUPLING imported in EPS format.

8   Change the PSQUALITY value to 100 and import the COUPLING .EPS file with:
    a) insertion point: 0,180
    b) scale factor: 100 – Fig. 12.3(c).

9   Import the COUPLING.EPS file again with:
    a) insertion point: 150,150
    b) scale factor: 200 – Fig. 12.3(d).

10   Set PSQUALITY to 0 and PSIN the COUPLING using:
    a) insertion point: 150,10
    b) scale factor: 140 – Fig. 12.3(e).

11   The two system variables used with this import format are:
    a) PSDRAG – displays the imported file and:
        0: only image boundary and file name displayed
        1: displays the rendered image
    b) PSQUALITY – determines the import display quality and:
        0: no image generated, only border and file name
        +ve: number of pixels per AutoCAD drawing unit. The greater the value, the better
               is the picture definition
        –ve: paths are displayed as outlines with no fill effect. This option is not applicable
               to our example.

12   *Questions.*
    a) can COUPLING.EPS be imported full size, i.e. circles at 100,120?
    b) what about layers?
    c) block or drawing?

Importing a BMP file

1   Menu bar with File–New–Start from Scratch–Metric–OK.

2   Menu bar with **Tools–Display Image–View** and:
    prompt   *Replay dialogue box*
    respond   1. File type: Bitmap (*.bmp)
                2. Look in: r14cust
                3. pick COUPLING then Open
    prompt   *Image Specifications dialogue box*
    respond   pick OK.

3   The COUPLING.bmp file will be displayed.

4   *Questions.*
    a) circle centres at 100,120?
    b) layers imported?
    c) can any of the display be erased or exploded?
    d) Windows taskbar displays Render?

5   Note that this is not a very good example to demonstrate how a BMP file can be imported
    into AutoCAD R14. The BMP import format is really meant for rendered images and not
    a 2D drawing as we have used.

Importing a solid ACIS file

1 Menu bar with File–New as before.

2 Menu bar with **Insert–ACIS Solid** and:
   prompt Select ACIS File dialogue box
   respond 1. Look in: r144cust
   2. pick SOLCOMP then Open.

3 The solid model is displayed in plan view.

4 Menu bar with **View–3D Viewpoint–SE Isometric** to display the model in 3D.

5 Menu bar with **View–Shade–16 Color Filled** and:
   a) model displayed as a black cube
   b) hole displayed in green.

6 What happened to the red cube?

Using other software packages with AutoCAD R14

It is very difficult for me to demonstrate how AutoCAD can be used with other software packages, as I do not know what packages are on your system. There is one package which all users should be able to access **Paint** – a Windows graphics application package. I will assume that this package is loaded into your system and we will investigate how to 'toggle' between AutoCAD and Paint.

1 Open the COUPLING **drawing** file and:
   a) cancel any floating toolbars
   b) turn the grid off
   c) cancel the screen menu if it is displayed.

2 Press the **Print Screen** button on your keyboard.

3 Pick Start from the Windows taskbar then **Programs–Accessories–Paint** and:
   prompt untitled Paint screen displayed
   respond menu bar with **Edit–Paste**.

4 The AutoCAD screen of the COUPLING drawing will be displayed, including title bar, scroll bars, etc.

5 Use the paint options to add some 'objects' to the screen.

6 Menu bar with **File–Save As** and:
   prompt Save As dialogue box
   respond 1. Save in: r14cust
   2. Save as type: 24-bit Bitmap
   3. File name: TEST
   4. pick Save.

7 Menu bar with File–Edit to return to AutoCAD.

8 Begin a new drawing from scratch with the Metric option and display toolbars as required.

9 At the command line enter **IMAGE <R>** and:

   prompt   *Image dialogue box*
   respond   **pick Attach**
   prompt   *Attach Image dialogue box*
   respond   1. Look in: r14cust
               2. pick TEST then Open
   prompt   *Attach Image dialogue box*
   respond   pick OK
   prompt   *Insertion point* and enter: **0,0 <R>**
   prompt   *Unit/Scale factor* and enter: **420 <R>**.

10 The inserted bitmap will be displayed as Fig. 12.4.

11 Can this image be altered?

12 *Note*: the IMAGE insertion command is really for rendered images and our example is not really suited to demonstrate the full potential of the command. We have only worked through the steps of inserting a file from another package into AutoCAD.

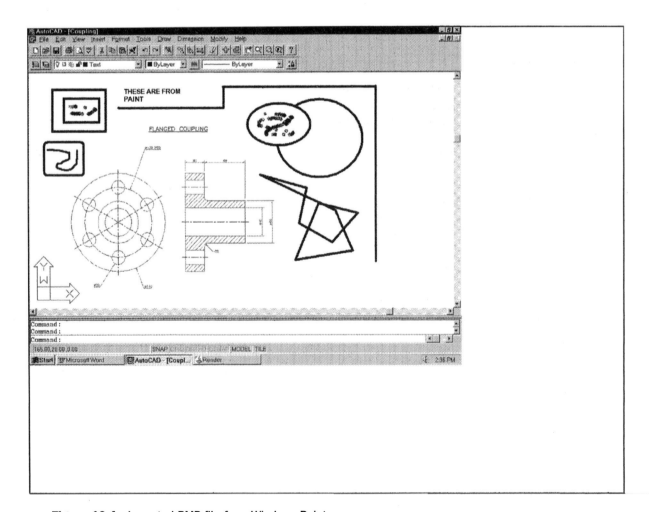

Figure 12.4 Imported BMP file from Windows Paint.

Inserting text files into AutoCAD

Most users will probably know how to import text files into an AutoCAD drawing. This chapter was to be only concerned with the import of graphical data into AutoCAD, but I thought that importing a text file would complete the 'picture'. We will therefore create a text file using Notepad, save it and import it into an existing drawing.

1 Open your COUPLING drawing, and add a rectangular border on layer 0 from 0,0 to 420,297. This border was erased for when exporting files. Move the drawing title to the lower part of the screen.

2 Activate Notepad from the Windows taskbar and enter the following lines of text:
   TECHNICAL REPORT ACADR14/AEFOR/17/80<R>
   <R>
   This report is an introduction to the work for the<R>
   JAMBOLY OIL COMPANY of MONYIA. This company is a new<R>
   multi-national with interests in both oil and gas<R>
   exploration, as well as in shipping and aviation.<R>
   The work which our company is undertaking is to<R>
   provide working drawings for a new refinery.<R>.

3 Menu bar with File–Save As and using the dialogue box:
   a) Look in: r14cust
   b) File name: REPORT_1.txt
   c) pick Save.

4 Minimize Notepad to return to AutoCAD.

5 Select the TEXT icon from the Draw toolbar and:
prompt	*Specify first corner* and enter: **190,290 <R>**
prompt	*Specify other corner* and enter: **410,215 <R>**
prompt	*Multiline Text Editor dialogue box*
respond	1. scroll at font name and pick Arial Black
	2. alter height to 5
	3. pick Import Text
	4. Look in: r14cust
	5. pick REPORT_1 then Open
prompt	*Text Editor with preview of text file*
respond	pick OK.

6 The text file will be imported into the drawing as Fig. 12.5.

7 Save if required but not as COUPLING.

8 This completes the last exercise in this chapter.

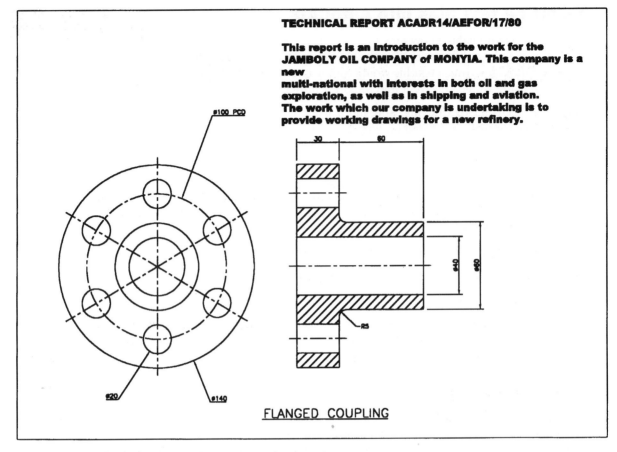

TECHNICAL REPORT ACADR14/AEFOR/17/80

This report is an introduction to the work for the JAMBOLY OIL COMPANY of MONYIA. This company is a new
multi-national with interests in both oil and gas exploration, as well as in shipping and aviation.
The work which our company is undertaking is to provide working drawings for a new refinery.

FLANGED COUPLING

Figure 12.5 Importing a text file into an AutoCAD drawing.

Summary

1 AutoCAD Release 14 supports many different file formats for export and import.

2 Export files can be used with other software packages.

3 Only a DXF import file gives a 'true definition' of a drawing as it imports all drawing geometry including layers, text and dimension styles, etc.

4 Many of the AutoCAD R14 import file formats are for rendered images which have not been considered in this book.

5 Text files can be imported into an AutoCAD drawing as paragraph text.

Assignments

No activity has been included with this chapter.

Chapter 13

Object linking and embedding

AutoCAD Release 14 allows drawing information to be 'transferred' between different Windows application packages and is called **OLE** – object linking and embedding. The procedure uses the CLIPBOARD package and allows the user to:

1 'Export' drawing information for 'import' to other applications.

2 Edit drawings while working in these other application packages.

3 Copy drawings into a current drawing.

To demonstrate OLE we will write a technical report containing AutoCAD drawings. This means that the user **must** have a word-processing package installed. I will use Microsoft Word, but the user can access any word processor.

OLE terminology

The basic terminology with OLE is:
a) Source document: the original information, e.g. an AutoCAD drawing.
b) Destination (compound) document: the package to which the source document is linked, e.g. a technical report which will contain the original AutoCAD drawing.

The original drawing

1 Start AutoCAD and open the drawing r14cust\COUPLING.

2 With layer 0 current, add a rectangular border from 0,0 to 420,297. This border was erased for the data exchange examples.

3 Move the complete drawing from 0,0 by @5,5.

4 Save the drawing layout as:
a) c:\r14cust\COUPLING
b) c:\r14cust\CPLING.

Linking a drawing

1 Coupling drawing with border displayed on screen?

2 Exit AutoCAD and using the Windows taskbar, activate Word.

3 Set the following:
  a) a text font and height – I used Courier with height 12
  b) full justification
  c) page set up to suit.

4 Enter the following lines of text:
  TECHNICAL REPORT ACADR14/AEFOR/17/80

  The couplings are for the JAMBOLY OIL of MONYIA
  and will be used in their new oil installation in the
  GONDOVIAN desert.
  The weather conditions in this environment are rather
  harsh, as the average day temperature is 38 degC and in
  the evening the temperature can drop to -50 degC. The
  prevailing wind is from the South East and can gust up
  to 150 km/hr.
  These conditions are the main reasons for using the new
  material **CASPUTIUM** in the construction of the
  couplings.

5 Ensure the cursor (|) is at the left of a 'new line'.

6 Menu bar with **Insert–Object** and:
  prompt   *Object dialogue box*
  respond   1. ensure **AutoCAD drawing** active in blue
        2. pick **Create from File** tab
  prompt   *Create from File dialogue box*
  respond   **pick Browse**
  prompt   *Browse dialogue box*
  respond   1. scroll and pick **CPLING drawing**
        2. pick OK
  prompt   *Browse dialogue box*
  with   **CPLING.dwg** as File name
  respond   1. activate **Link to File** (i.e. tick)
        2. pick OK.

7 AutoCAD R14 will be 'started' with the CPLING drawing displayed.

8 Exit AutoCAD to return to Word and the CPLING drawing will be displayed below the
  entered text, i.e. it is linked to the text document – Fig. 13.1.

9 Menu bar with **File–Save As** and:
  prompt   *Save As dialogue box*
  respond   1. Save in: **r14cust**
        2. File name: **REPORT**
        3. pick Save.

10 Before leaving Word, menu bar with **Tools–Options** and:
  prompt   *Options dialogue box*
  respond   1. activate the **General** tab
        2. ensure **Update Automatic Links at Open** is activated, i.e. tick in box
        3. pick OK.

11 Exit Word.

TECHNICAL REPORT ACADR14/AEFOR/17/80

The couplings are for the JAMBOLY OIL COMPANY of MONYIA and will be used in their new oil installation in the GONDOVIAN desert.
The weather conditions in this environment are rather harsh, as the average day temperature is 38 degC and in the evening the temperature can drop to -50 degC. The prevailing wind is from the South East and can gust up to 150 km/hr.
These conditions are the main reasons for using the new material CASPUTIUM in the construction of the couplings.

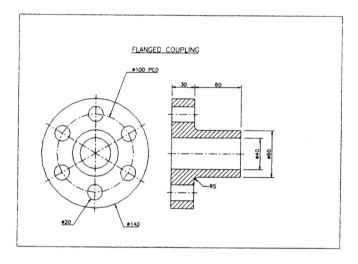

Figure 13.1   Document with original linked drawing.

Editing a linked drawing

Source documents (AutoCAD drawings) which have been linked to other packages can be edited from:
a) within the destination (compound) document, i.e. from Word
b) the original AutoCAD drawing.

Editing from Word

1   Ensure AutoCAD is not 'opened'.

2   Start Word and open **c:\r14cust\REPORT** to display the technical report and the original CPLING drawing.

3   *a*) move the cursor (I) into the 'drawing area'
    b) double left-click in the drawing area.

4   AutoCAD will be 'opened' and the CPLING drawing displayed.

5   Alter the drawing to make the right view an outside elevation:
    a) no hatching
    b) hidden lines
    c) new 'holes'
    d) you should have the ability to complete this!

6   Menu bar with **File–Save** to update CPLING.

7   Menu bar with **File–Exit** to return to Word with the drawing displayed with these alterations – Fig. 13.2.

8   Menu bar with **File–Save** to update REPORT, then exit Word.

Editing from AutoCAD

1   Start AutoCAD and open drawing **c:\r14cust\CPLING**.

2   Drawing displayed with previous alterations?

3   Using the **STRETCH** command (icon?), stretch the right end of the right view (six lines and three dimensions) by @**51.25,0**.

4   *a*) menu bar with **File–Save** to update CPLING
    b) menu bar with **File–Exit**.

5   Start Word and open document **c:\r14cust\REPORT**.

6   The report will be 'opened' but will 'jump' to AutoCAD to display the stretched CPLING drawing – this is because the CPLING drawing has been modified since it was last saved in the Word document and the Update Automatic Links was activated.

7   Exit AutoCAD and the document will display the stretched CPLING drawing – Fig. 13.3.

8   Menu bar with File–Save to update REPORT then exit Word.

TECHNICAL REPORT ACADR14/AEFOR/17/80

The couplings are for the JAMBOLY OIL COMPANY of MONYIA and will be used in their new oil installation in the GONDOVIAN desert.
The weather conditions in this environment are rather harsh, as the average day temperature is 38 degC and in the evening the temperature can drop to -50 degC. The prevailing wind is from the South East and can gust up to 150 km/hr.
These conditions are the main reasons for using the new material CASPUTIUM in the construction of the couplings.

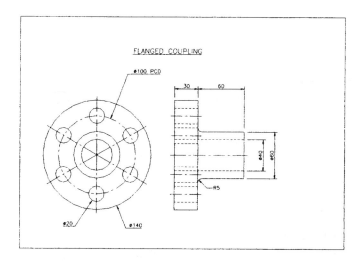

Figure 13.2   REPORT with modified CPLING drawing.

TECHNICAL REPORT ACADR14/AEFOR/17/80

The couplings are for the JAMBOLY OIL COMPANY of MONYIA and will be used in their new oil installation in the GONDOVIAN desert.
The weather conditions in this environment are rather harsh, as the average day temperature is 38 degC and in the evening the temperature can drop to -50 degC. The prevailing wind is from the South East and can gust up to 150 km/hr.
These conditions are the main reasons for using the new material CASPUTIUM in the construction of the couplings.

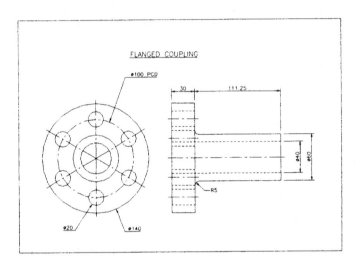

Figure 13.3 REPORT with 'stretched' CPLING drawing.

Embedding a drawing

Embedding a source drawing into a compound (destination) document is very similar to linking, and we will discuss the differences later in this chapter. To demonstrate the process, we will create a new drawing and embed it into our report, so:

1 Start AutoCAD R14 and open STDA3 template file.

2 Refer to Fig. 13.4 and draw the component as shown. Use layers correctly and add any refinements of your choice.

3 Save the completed drawing as **c:\r14cust\ENDCAP**.

4 Menu bar with **Edit–Copy** and:
   prompt   *Select objects*
   enter   **ALL <R>**
   prompt   *?? found* then *Select objects*
   respond **right-click**.

5 Activate Word from the Windows taskbar and open the REPORT document from the c:\r14cust folder. The report should be displayed with the stretched component as Fig. 13.3.

6 Add the following text 'below' the drawing:

   The flanged couplings require a special 'end cap' made from **MACFARALUMIN**, manufactured to the following sizes:

7 Enter a <RETURN> after the text then select from the menu bar **Edit–Paste**.

8 The endcap drawing will be pasted into the report as Fig. 13.5.

9 Save the report as **c:\r14cust\REPORT**.

10 Exit Word to return to AutoCAD and exit AutoCAD. Save changes?

Editing an embedded drawing

Embedded drawings **can only be edited from the document in which they are embedded**, so:

1 Start Word and open the REPORT document.

2 Move the cursor into the endcap drawing and double left-click.

3 AutoCAD will be 'opened' with the endcap drawing displayed.

4 Modify the drawing as follows:
   a) erase all text and dimensions
   b) erase the hidden detail in the left view
   c) replace the six circles in the right view hexagons, inscribed in a circle of radius 10 – easy?
   d) complete the left view by adding 'bolt heads' of height 8 to the left of the left vertical line using the six hexagons as reference
   e) erase all centre lines.

5 Menu bar with **File** and:
   prompt   *pull down menu*
   with   a new item – **Update Microsoft Word**
   respond **pick this new item**
   then   exit AutoCAD to return to Word.

6 The document will display the endcap drawing with these alterations – Fig. 13.6.

7 Save the document as c:\r14cust\REPORT then exit Word.

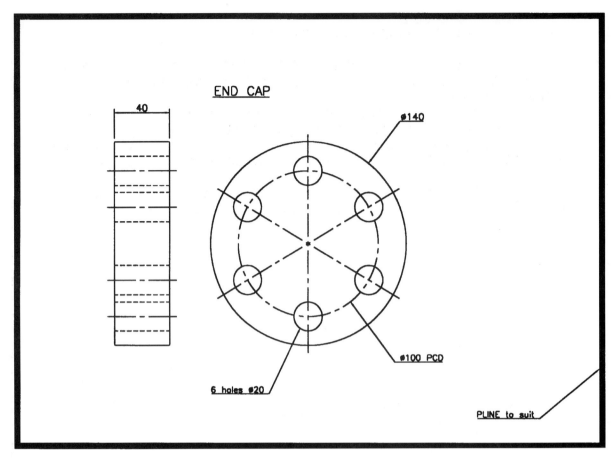

Figure 13.4 Endcap drawing for embedding.

Linking vs embedding

Linking and embedding are similar processes. The correct terminology is Object Linking and Embedding (OLE) and is a Windows facility. Both options allow an AutoCAD drawing to be 'imported' into Windows application software – in our case Word, a word processor. The difference between the options is in the manner in which the drawing data is stored.

Linking: This actually makes a link between the original (**source**) drawing and the destination (**compound**) document. The link informs the destination document where the source drawing data can be found.

a) The linked drawing can be edited from the destination document or from AutoCAD and both options will update the source document.

b) Linking is useful if the destination document has to have automatic updates when the source drawing is altered.

c) Only complete named drawings can be linked.

d) Linking can be 'likened' to external references.

Embedding: An embedded drawing is actually copied into the destination document and is not directly associated with the source drawing.

a) An embedded drawing can only be edited from the actual document and the source drawing is unaltered

b) embedding should be used when the source drawing is to remain unaltered and the drawing in the document is to be modified

c) Individual objects, e.g. a window effect, can be embedded.

TECHNICAL REPORT ACADR14/AEFOR/17/80

The couplings are for the JAMBOLY OIL COMPANY of MONYIA and will be used in their new oil installation in the GONDOVIAN desert.
The weather conditions in this environment are rather harsh, as the average day temperature is 38 degC and in the evening the temperature can drop to -50 degC. The prevailing wind is from the South East and can gust up to 150 km/hr.
These conditions are the main reasons for using the new material **CASPUTIUM** in the construction of the couplings.

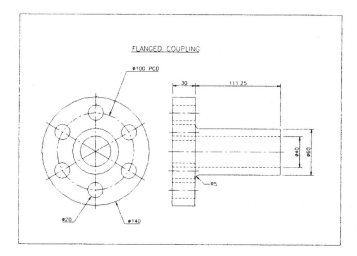

The flanged coupling requires a special 'end cap' made from **MACFARALUMIN**, manufactured to the following sizes:

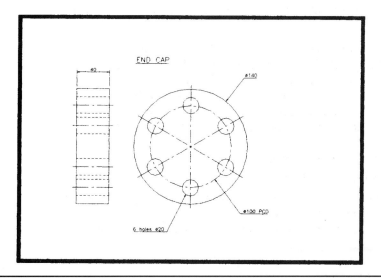

Figure 13.5 Endcap drawing embedded into the report.

TECHNICAL REPORT ACADR14/AEFOR/17/80

The couplings are for the JAMBOLY OIL COMPANY of MONYIA
and will be used in their new oil installation in the
GONDOVIAN desert.
The weather conditions in this environment are rather
harsh, as the average day temperature is 38 degC and in
the evening the temperature can drop to -50 degC. The
prevailing wind is from the South East and can gust up
to 150 km/hr.
These conditions are the main reasons for using the new
material   CASPUTIUM   in   the   construction   of   the
couplings.

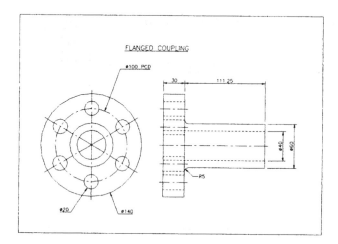

The flanged coupling requires a special 'end cap' made
from MACFARALUMIN, manufactured to the following sizes:

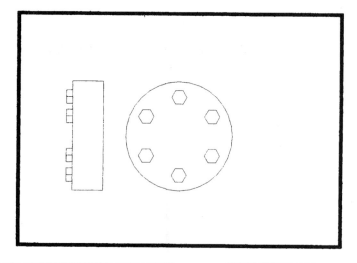

Figure 13.6   Altered endcap drawing.

Task

Start AutoCAD and:
a) open c:\r14cust\ENDCAP: original with circles, i.e. not affected from the embedding exercise
b) open c:\r14cust\CPLING: the 'stretched' component should be displayed, i.e. the linked alterations have affected the original drawing.

Copying drawings

AutoCAD objects can be copied to the **Clipboard** package for insertion into other software or for insertion into another AutoCAD drawing. To demonstrate the process:

1 Start AutoCAD R14 and open drawing COUPLING. This was the original drawing saved at the start of the chapter with a border added.

2 Erase all dimensions.

3 From the Standard toolbar select the COPY TO CLIPBOARD icon and:
prompt   '_copyclip
then   Select objects
respond **window the complete drawing then right-click.**

4 Activate Programs–Accessories–Clipboard Viewer with Start from the Windows taskbar and:
a) Clipboard Viewer screen displayed
b) double left-click on page 1 tab
c) Coupling 'drawing' displayed?
d) Cancel Clipboard to return to AutoCAD.

5 Open the ENDCAP drawing (No to any changes) and:
a) erase all dimensions
b) move the component as far to the right as possible, but still within the polyline border.

6 From the Standard toolbar select the Paste to Clipboard icon and:
prompt   '_pasteclip
then   Insertion point and enter: **20,20 <R>**
prompt   X scale factor and enter: **0.35 <R>**
prompt   Y scale factor and enter: **0.65 <R>**
prompt   Rotation angle and enter: **0 <R>**.

7 The Coupling drawing will be inserted from Clipboard into the current drawing – Fig. 13.7.

8 *Questions*:
a) is the inserted drawing a block?
b) can it be exploded?
c) are the layers correct for the inserted object?

9 Save the drawing if required, as this completes the OLE exercise.

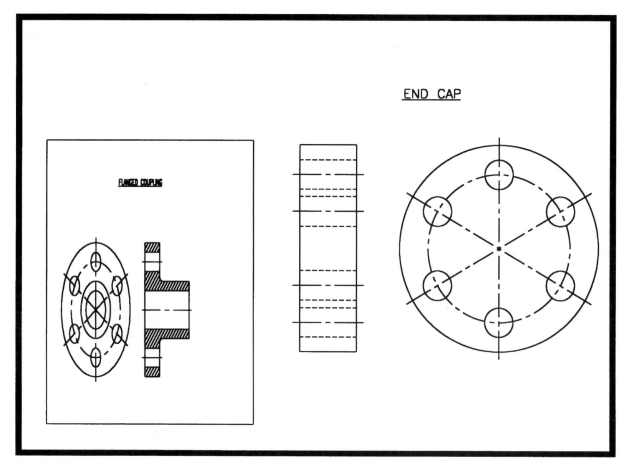

Figure 13.7 Copied drawing pasted from Clipboard into an existing drawing.

Summary

1   AutoCAD objects/drawings can be 'linked' or 'embedded' to other Windows application software.

2   AutoCAD drawings can be copied into Clipboard, which can be used as a 'transfer' medium between packages.

3   The differences between linking and embedding are:

Linking	*Embedding*
Complete drawings	Selected objects
Edited from AutoCAD or document	Edited from document
Source drawing altered	Source drawing unaltered.

4   When linking an AutoCAD drawing to another package, the 'linking' procedure of the package must be followed.

5   Embedding is the easier process.

6   Readers who have linked AutoCAD drawings using LT/R13 will have realized that the process is slightly different with R14.

7   This chapter has been an introduction to OLE. The concept is very powerful and all the options have not been covered.

Chapter 14

Linking with a database

AutoCAD R14 has a 'built-in' database. If the full installation is loaded from the CD-ROM, the user has access to several 'named' database packages. In this chapter we will demonstrate the process using an AutoCAD drawing which has associated database files with the extension **dbf**, i.e. they have been created using Dbase III.

Notes

1 It must be appreciated that the exercise is only an introduction to linking AutoCAD drawings with a database. The topic is more involved than could be covered in a single chapter.

2 The user will encounter several new dialogue boxes. These will not be displayed.

Terminology

Databases have their own unique terminology and the following are some of the 'main' terms encountered when linking AutoCAD drawings to a database:
a) ASE: AutoCAD SQL Extension
b) SQL: Structured Query Language
c) Environment: a database management system (DBMS)
d) Catalog: a directory path name to locate schemas
e) Schema: a catalog subdirectory of tables
f) Table: a group of related objects organized in rows and columns.

The drawing

To investigate AutoCAD and database, a drawing is required which has attribute information in a database format, so:

1 Start AutoCAD and open **c:\AutoCAD R14\Sample\asesmp** to display a plan view of an office layout.

2 Zoom in on the upper right area of the plan to display room 109 similar to Fig. 14.1.

3 The zoomed area displays the following blocks:
name	*colour*
CHAIRS	cyan
FURNITURE	green
CPU	magenta
PARTITION	blue
WALLS	black.

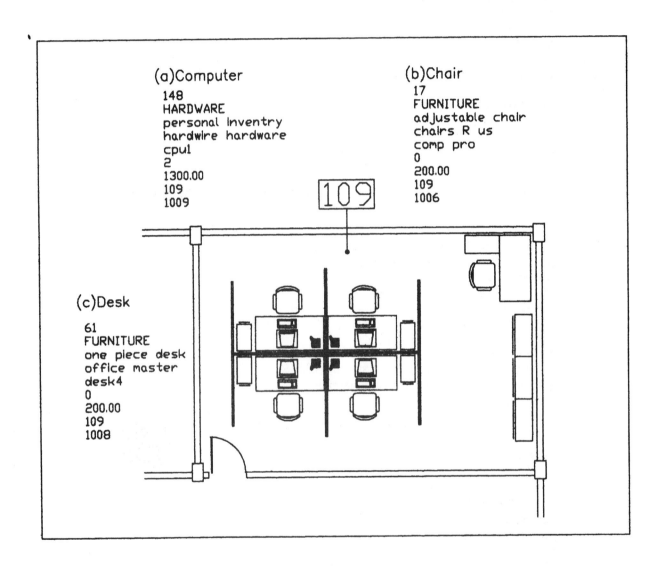

Figure 14.1 Zoomed area of ASESMP with three displayable attributes.

Activating the DBMS

1 Display the External Database toolbar and select the ADMINISTRATION icon and:

    prompt   *Administration dialogue box*
    with   DB3, ODBC, ORACLE7
    and   1. DB3 highlighted in blue
                2. Environment active (black dot)
    respond **\<RETURN\>**
    prompt   *Catalog active*
    respond **pick ASE (blue) then \<R\>**
    prompt   *Schema active*
    respond **pick DB3SAMPLE then \<R\>**
    prompt   *Table active*
    and   COMPUTER,EMPLOYEE,INVENTRY
    respond **pick INVENTRY**
    and   Database Object Setting:
                DB3.ASE.DB3SAMPLE.INVENTRY
    respond **pick OK.**

2 We have investigated how to 'set' the various parts of the database, but have not yet been 'connected' to it.

3 Menu bar with **Tools–External Database–Administration** and:
    prompt   *Administration dialogue box – as left in step 1?*
    respond **pick Environment then Connect**
    prompt   *Connect to Environment dialogue box*
    respond **pick OK** – password not needed
    prompt   *Table active*
    respond 1. pick INVENTRY
                2. scroll at Link Path Name at right
                3. pick INV
    and   Database Object Setting:
                DB3.ASE.DB3SAMPLE.INVENTRY(INV)
    respond **pick OK.**

4 Select the ROWS icon from the External Database toolbar and:
    prompt   *Rows dialogue box*
    with   Environment: DB3
                Catalog: ASE
                Schema: DB3SAMPLE
                Table: INVENTRY
                Link Path Name: INV
    respond **pick Graphical\<**
    prompt   *Select object*
    respond **pick any magenta computer symbol**
    prompt   *Rows dialogue box*
    with   data about the object selected
    respond **pick Make DA**
    prompt   *Make Displayable Attribute dialogue box*
    respond **pick Add–All→** to transfer Table to DA Column
    then   **pick OK**
    prompt   *Left point*
    respond **pick a point to suit**
    prompt   *Rows dialogue box*
    respond **pick OK.**

5 The attribute data for the selected object will be displayed similar to Fig. (a).

6 Menu bar with **Tools–External Database–Rows** and:

    prompt   *Rows dialogue box*

    respond   1. pick Graphical<

                  2. pick any cyan chair

                  3. pick Make DA

                  4. ensure DA Column has all the Table Column items

                  5. pick OK

                  6. pick a left point to suit

                  7. pick OK from Rows dialogue box.

7 Chair data displayed as Fig. (b).

8 *Task.*

    Using the Rows dialogue box, display the attribute data for any green desk object – Fig. (c).

9 The drawing should be similar to Fig. 14.1.

10 Save and exit.

Note

This completes the simple exercise into linking AutoCAD drawings with a database. The concept is very powerful and is really beyond the scope of this book. The chapter has been included to allow the user a 'glimpse into another use for AutoCAD'.

Chapter 15

And finally ...

Hopefully by the time you read this chapter you will have completed all the exercises and assignments and are now reasonably knowledgeable about the more advanced topics available in AutoCAD Release 14.

As with all my previous publications, I welcome any comments from readers about:

• improvements that could be made the book
• other areas of AutoCAD which reader's require material
• new drawing ideas.

I hope you have enjoyed learning from this book.

Bob McFarlane
Bellshill, Scotland

ACTIVITY 1(a)
1. Draw the lorry icon as fig(a)
2. Define 5 attributes using data from fig(b)
3. Icon with tags similar to fig(c)
4. Make a block of the icon and tags, the block name being LORRY
5. Use the attribute data from fig(d) to insert the LORRY block seven times.

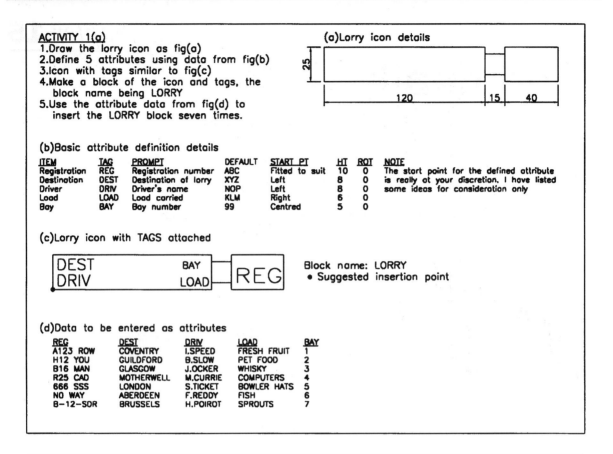

(a)Lorry icon details

(b)Basic attribute definition details

ITEM	TAG	PROMPT	DEFAULT	START PT	HT	ROT	NOTE
Registration	REG	Registration number	ABC	Fitted to suit	10	0	The start point for the defined attribute
Destination	DEST	Destination of lorry	XYZ	Left	8	0	is really at your discretion. I have listed
Driver	DRIV	Driver's name	NOP	Left	8	0	some ideas for consideration only
Load	LOAD	Load carried	KLM	Right	6	0	
Bay	BAY	Bay number	99	Centred	5	0	

(c)Lorry icon with TAGS attached

Block name: LORRY
• Suggested insertion point

(d)Data to be entered as attributes

REG	DEST	DRIV	LOAD	BAY
A123 ROW	COVENTRY	I.SPEED	FRESH FRUIT	1
H12 YOU	GUILDFORD	B.SLOW	PET FOOD	2
B16 MAN	GLASGOW	J.OCKER	WHISKY	3
R25 CAD	MOTHERWELL	M.CURRIE	COMPUTERS	4
666 SSS	LONDON	S.TICKET	BOWLER HATS	5
NO WAY	ABERDEEN	F.REDDY	FISH	6
B-12-SDR	BRUSSELS	H.POIROT	SPROUTS	7

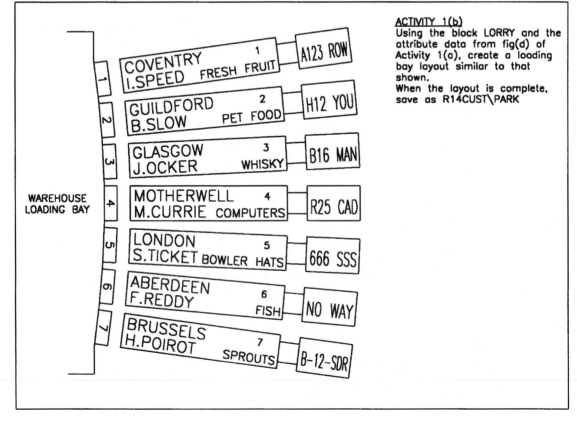

ACTIVITY 1(b)
Using the block LORRY and the attribute data from fig(d) of Activity 1(a), create a loading bay layout similar to that shown.
When the layout is complete, save as R14CUST\PARK

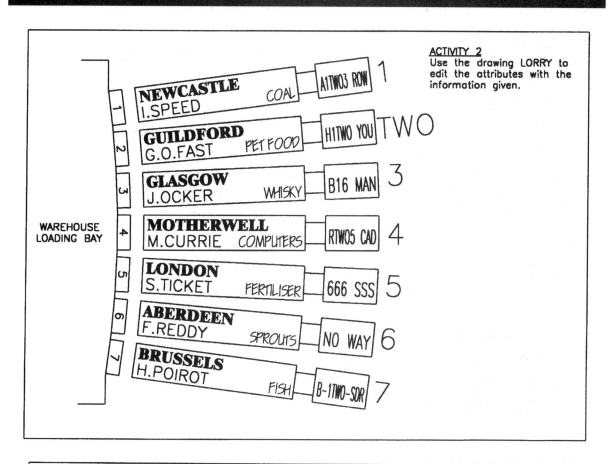

ACTIVITY 2
Use the drawing LORRY to
edit the attributes with the
information given.

Note: ATTDISP set to 0 for clarity.

ACTIVITY 3
The attribute extract files for the
warehouse loading bay:
File A; PARKCDF.txt
File B: PARKSDF.TXT
File C: modified PARKSDF file

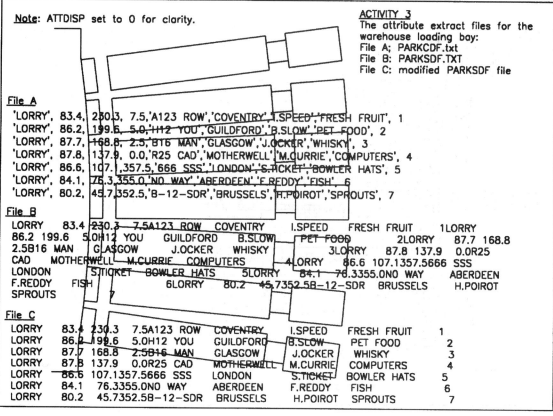

File A
'LORRY', 83.4, 250.3, 7.5,'A123 ROW','COVENTRY','I.SPEED','FRESH FRUIT', 1
'LORRY', 86.2, 199.6, 5.0,'H12 YOU','GUILDFORD','B.SLOW','PET FOOD', 2
'LORRY', 87.7, 168.8, 2.5,'B16 MAN','GLASGOW','J.OCKER','WHISKY', 3
'LORRY', 87.8, 137.9, 0.0,'R25 CAD','MOTHERWELL','M.CURRIE','COMPUTERS', 4
'LORRY', 86.6, 107.1,357.5,'666 SSS','LONDON','S.TICKET','BOWLER HATS', 5
'LORRY', 84.1, 76.3,355.0,'NO WAY','ABERDEEN','F.REDDY','FISH', 6
'LORRY', 80.2, 45.7,352.5,'B-12-SDR','BRUSSELS','H.POIROT','SPROUTS', 7

File B
LORRY   83.4 230.3   7.5A123 ROW   COVENTRY   I.SPEED   FRESH FRUIT   1LORRY
86.2 199.6   5.0H12 YOU   GUILDFORD   B.SLOW   PET FOOD   2LORRY   87.7 168.8
2.5B16 MAN   GLASGOW   J.OCKER   WHISKY   3LORRY   87.8 137.9   0.0R25
CAD   MOTHERWELL   M.CURRIE   COMPUTERS   4LORRY   86.6 107.1357.5666 SSS
LONDON   S.TICKET   BOWLER HATS   5LORRY   84.1 76.3355.0NO WAY   ABERDEEN
F.REDDY   FISH   6LORRY   80.2 45.7352.5B-12-SDR   BRUSSELS   H.POIROT
SPROUTS   7

File C
LORRY   83.4 230.3   7.5A123 ROW   COVENTRY   I.SPEED   FRESH FRUIT   1
LORRY   86.2 199.6   5.0H12 YOU   GUILDFORD   B.SLOW   PET FOOD   2
LORRY   87.7 168.8   2.5B16 MAN   GLASGOW   J.OCKER   WHISKY   3
LORRY   87.8 137.9   0.0R25 CAD   MOTHERWELL   M.CURRIE   COMPUTERS   4
LORRY   86.6 107.1357.5666 SSS   LONDON   S.TICKET   BOWLER HATS   5
LORRY   84.1 76.3355.0NO WAY   ABERDEEN   F.REDDY   FISH   6
LORRY   80.2 45.7352.5B-12-SDR   BRUSSELS   H.POIROT   SPROUTS   7

ACTIVITY 4
1. Draw the original NUT
2. Save as NUT – remember BASE
3. Draw the three circles
4. Attached the XREF NUT then polar array
5. Save layout as XREFACT
6. Open drawing NUT and modify as shown
7. Save NUT
8. Open drawing XREFACT
9. Modifications displayed?

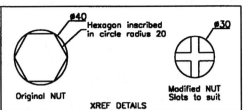

Ø40 Hexagon inscribed
in circle radius 20

Ø30

Original NUT

Modified NUT
Slots to suit

XREF DETAILS

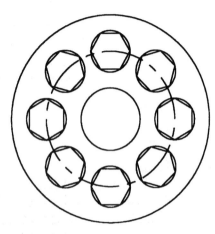

Layout with original NUT Xref

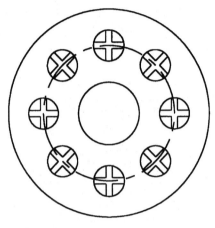

Layout with modifed NUT Xref

ACTIVITY 5
Using the created simple, complex and multiline linetypes,
create a sports stadium to your own design.
Optimise the linetype variables.
Save as R14CUST\STADIUM

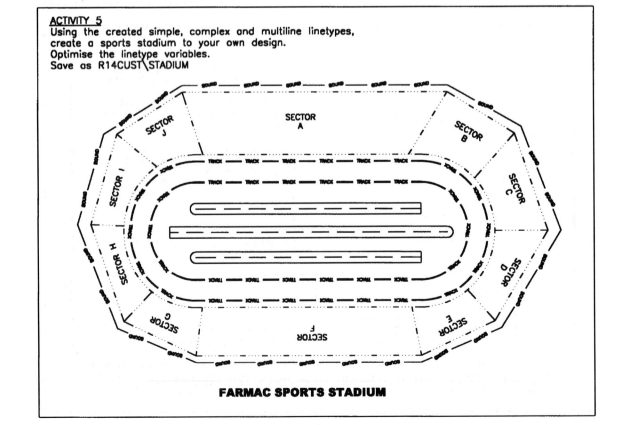

FARMAC SPORTS STADIUM

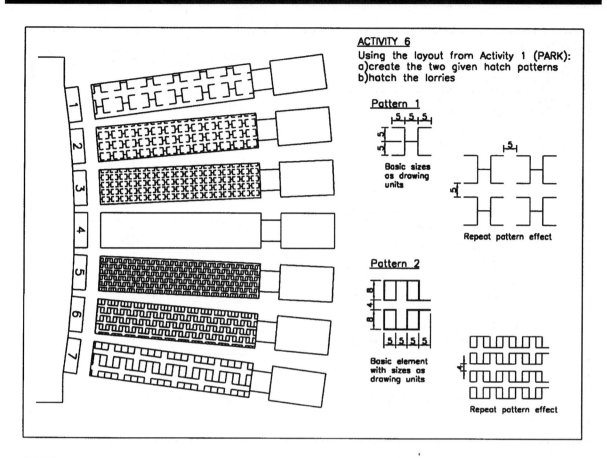

ACTIVITY 6

Using the layout from Activity 1 (PARK):
a)create the two given hatch patterns
b)hatch the lorries

Pattern 1

Basic sizes
as drawing
units

Repeat pattern effect

Pattern 2

Basic element
with sizes as
drawing units

Repeat pattern effect

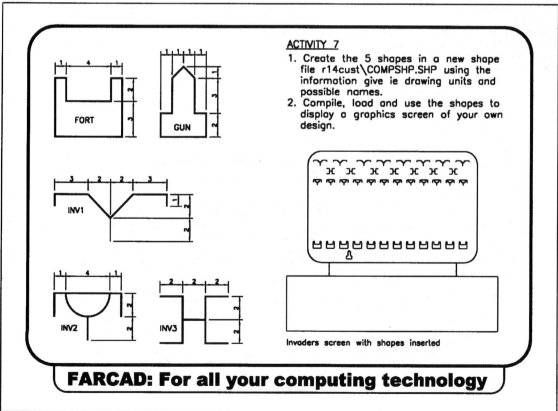

ACTIVITY 7

1. Create the 5 shapes in a new shape file r14cust\COMPSHP.SHP using the information give ie drawing units and possible names.
2. Compile, load and use the shapes to display a graphics screen of your own design.

FORT

GUN

INV1

INV2

INV3

Invaders screen with shapes inserted

FARCAD: For all your computing technology

ACTIVITY 8

1. Use the blocks from Fig.11.13 and the modified MYMENU.MNU to draw the electrical circuit
2. Extract the attribute information in CDF and SDF formats using a template file.

ELECTRICAL CIRCUIT

Imported SDF attribute extraction file:

RES	R20
RES	R1
NPN	Q1
DIO	D1
RES	R3
RES	R5
NPN	Q3
NPN	Q5
PNP	Q4
PNP	Q6
PNP	Q10
RES	R8
RES	R9
CAP	C1
RES	R6
RES	R4
PNP	Q2
DIO	D2
RES	R2
NPN	Q7
NPN	Q8
PNP	Q9
RES	R7
RES	R10
NPN	Q11
PNP	Q12
PNP	Q13
NPN	Q14
RES	R13
CAP	C2
NPN	Q15
NPN	Q16
RES	R15
RES	R17
RES	R12
DIO	D4
RES	R16
RES	R18
PNP	Q17
RES	R11
RES	R14
DIO	D4
CAP	C3
DIO	D6
RES	R19
CAP	C4
DIO	D5

ACTIVITY 9

Digitise the drawing as
much as possible, and
use your expertise to
complete it.

Pt2

Digitised points:
Pt1: 0,0
Pt2: 210,290

THE TYGER

Tyger! Tyger! burning bright
In the forests of the night
What immortal hand or eye
Could frame thy fearful symmetry?

In what distant deeps or skies
Burnt the fire of thine eye?
On what wings dare he aspire?
What the hand dare seize the fire?

And what shoulder, and what art?
Could twist the sinews of thy heart?
And when thy heart began to beat
What dread hand? and what dread feet?

What the hammer? what the chain?
In what furnace was thy brain?
What the anvil? what dread grasp
Dare its deadly terrors clasp?

When the stars threw down their spears
And watered heaven with their tears,
Did he smile his work to see?
Did he who made the Lamb make thee?

Tyger! Tyger! burning bright
In the forests of the night,
What immortal hand or eye,
Dare frame thy fearful symmetry?

Courtesy of:

WILLIAM BLAKE 1757-1827

Pt1

Index